The Making Of Ireland
Landscapes In Geology

by
Michael Williams and
David Harper

IMMEL
Publishing

The Making Of Ireland is published by Immel Publishing
© 1999 Michael Williams and David Harper

The right of Michael Williams and David Harper to be identified as the authors of this work has been asserted in accordance with the Copyright Design and Patents Act 1988, sections 77 and 78.

Design & Typesetting: Justin King
Printed by MPG Books, Bodmin, Cornwall.

All rights reserved. No part of this publication may be reproduced, stored in a retrieval syatem or transmitted in any form or by any means, electronic, electrostatic, magnetic tape, mechanical, photocopying, recording or otherwise, without permission in writing from the publishers.

Cataloguing in Publication Data
A CIP catalogue record for this book is available from the British Library.

ISBN 1 898162 069

Immel Publishing Limited
14, Dover Street, London W1X 3PH
Tel: 0171 491 1799, Fax: 0171 493 5524

Contents

Preface

Chapter 1. How to Create a Country.   2
    Plate Tectonics
    Classes of Rock
    Geological Time

Chapter 2. In the Beginning.   10
    The Oldest Irish Rocks
    The Origins of Connemara
    Life in the Precambrian

Chapter 3. The Birth and Death of an Ocean.   18
    Ordovician Volcanoes
    A Silurian Flooding

Chapter 4. The Rise and Fall of the Mountains.   28
    The Great Caledonian Mountains
    The Mountains Crumble
    The Arrival of Life on Land

Chapter 5. Tropical Climates.   34
    Diving in the Carboniferous Sea
    Bye Bye Bahamas
    The Mississippi comes to Clare
    A Gentle Orogeny
    The legacy of the Carboniferous

Chapter 6. The Atlantic Opens.   42
    Salty Waters
    An Extinction of Species
    The Time of the Dinosaurs
    The Chalk Sea
    Did Meteorites Kill the Dinosaurs?
    A New Ocean
    Oil and Gas

Chapter 7. The Ice Man Cometh.   54
    The Irish Glaciers
    Causes of Ice Ages
    Irish Animals in the Pleistocene
    The Earliest Irish

Trail Guide 1: South-West Donegal   65

Trail Guide 2: Connemara   69

Trail Guide 3: Clare   73

Trail Guide 4: Dingle   77

Trail Guide 5: Wexford   81

Trail Guide 6: The Antrim Coast   85

A glossary of some geological terms   92

References   96

Preface

This book leads you on a journey through deep time. It is a book about the geological history of Ireland, and we will see its landscapes and seascapes evolve from the distant past to the present day. Fragments of the country began as disparate blocks of the planet's crust, which became joined together until, finally, the modern landscape was sculpted by the last glaciation.

The book is written for people who have no previous knowledge of earth sciences and basic principles are clarified throughout the text. There is also a glossary of basic geological terms at the end of the text. We have tried to ensure that all the terms we use are explained so that they are clear to the lay reader. It is not intended to be a rigorous analysis of Irish Geology. For those who wish to delve further into geological details there is a list of starting references. For those of you who think of rocks as hard, grey and rather boring, we hope to show that by learning to read them, you can discover something about the extraordinary set of circumstances that led to the making of Ireland. Trail guides are included to show that the rocks in various parts of the country contain visible evidence of the ways in which they were formed.

Geological ideas are always evolving in an attempt to explain the workings of the earth. The ideas presented in the book are therefore stepping stones to a full understanding of the evolution of Ireland. Geological discoveries are constantly being made, each one modifying, to a greater or lesser extent, the nature of the science. We do not have all the answers and it would be a dull world indeed if we did.

We depend for our existence on the atmosphere, the waters of the planet and the crust beneath our feet. As well as protecting the atmosphere and keeping our oceans and rivers free of pollution, we must also understand how our Earth works. We hope this book will help you some way towards that understanding.

We thank both Ian Mitchell and Kevin Fitzpatrick for reading parts of early versions of the manuscript. We are grateful to Secker & Warburg for permission to quote from *Irish Journal* by Heinrich Böll and to J.M. Dent for *Collected Poems (1995)* by R.S. Thomas.

We dedicate this book to the memory of Tony Whilde whose enthusiasm for natural history was immense.

Michael Williams & David Harper, Galway, Ireland, 1995.

How to Create a Country

Plate Tectonics

Classes of Rock

Geological Time

CHAPTER ONE

How to Create a Country

"The earth has a mass of 5.97 x 10^{24} kilograms,...... a big number and one that really matters because that is all the matter we have got."
David Bellamy, Botanic Man (1978)

Ireland contains some of the most impressive and unspoilt scenery in western Europe. From the conical white quartzite mountains of Donegal in the north to the ragged clutching fingers of the Skellig Rocks in the south, the constant changes of light and shade alter these vistas moment by moment. This country was forged on the anvil of time in a way which reflects the constantly changing patterns and configurations in the history of the planet.

This book attempts to guide you through the geological evolution of this part of the earth's crust. Piece by piece we will see how the jigsaw was fitted together over millions of years. The rocks of Ireland contain the clues to its past; every piece of rock acts as a sort of fossilized video film, a record of the events which have contributed to the creation of the country. These events include the birth and death of mighty oceans, great desert dunes sweeping over a parched landscape, violent explosive eruptions of volcanoes and the eventual first tentative steps of humans onto its shores.

Ireland achieved its present shape only recently in geological terms. To understand how this came about we have first to appreciate how the earth works. In broad terms its structure is relatively simple. The outer skin of the globe consists of a hardened rocky crust, which varies in thickness from about 5 to 40 kilometres. The height of the crust varies from nearly 9 kilometres above sea level on Mount Everest to over 4.5 kilometres below sea level in parts of the northern Pacific Ocean. Yet, even starting at this low point in the Pacific it would still require a journey of over 6000 kilometres to reach the earth's centre. So we can see that the crust on which we live is only a small fraction of the volume of the planet.

This globe contains concentric shells of material with differing properties. The shells that mainly concern us are the crust and the underlying upper mantle. The upper mantle is so hot and under such pressure that it is capable of flowing. It could be thought of as semi-molten rock whereas the overlying crust is far more rigid and, of course, cooler.

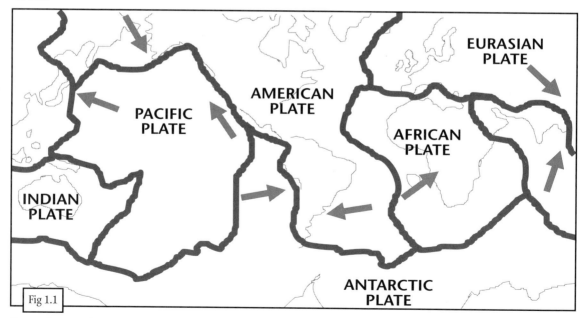

Fig 1.1. The major plates of the earth's crust with arrows showing their approximate relative motions at present. There are a number of smaller plates not shown here. Ireland is placed well away from any plate boundary which explains why it suffers from only minor earthquakes and no volcanic eruptions. Notice how Iceland sits directly on a spreading ridge and is therefore a volcanic centre.

Plate Tectonics

Many scientists believe that the evolution of the crust of the earth can be explained by a theory called Plate Tectonics, which is partially based on the work of Alfred Wegner, who in 1912 suggested that the continents had moved from their original positions. He called this process Continental Drift. The Plate Tectonic theory, which is less than 30 years old, explains how oceans are created and destroyed, and explains how mountain ranges are built. It also helps us to understand the location and causes of many earthquakes and volcanoes.

This theory suggests that the crust of the earth is divided into a number of sectors, or plates, each is moving with respect to its neighbour, as shown in Figure 1.1. One could think of these rigid plates as floating on the less rigid mantle of the earth. These days this can be proved by using satellite-based systems which measure the movement of the continents year by year. A very average rate of movement might be 2 centimetres per year. The surface area of the earth remains more or less constant, but, if these plates are not all moving in the same direction and at the same speed, we are left with a problem of space on the global surface. If two plates are moving apart, something must fill in the space created between them and if they are moving towards each other, then one plate must somehow give way to the other.

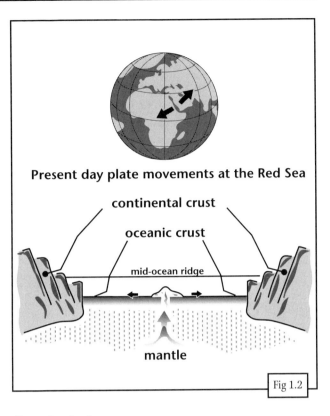

Two types of crust make up the skin of the earth: oceanic and continental. Oceanic crust is relatively thin, about 5-10 kilometres and is made of dense material containing heavy minerals, that is minerals containing heavy elements such as iron and magnesium. Continental crust is normally thicker (20 - 40 kilometres) but less dense. It is made up of much lighter material containing minerals such as quartz. If a continent begins to split apart, as in Figure 1.2, the space created is filled by molten rock injected from the mantle. This molten rock is under pressure from the weight of overlying crust and escapes through the faults along which the continent is splitting. When it reaches the surface in the form of volcanic outpourings, it hardens and forms new crust of the heavy oceanic type. As the pieces of the continent move farther apart, more and more new crust is formed in the split. This is happening today in the East African Rift valley for example. Here Africa and Arabia are moving apart. The space caused by the splitting of an original continent has been filled by new oceanic crust which forms the floor to the Red Sea. Eventually as this gap widens and subsides it will become an ocean. New material is added to oceanic crust along a spreading centre where a line of submarine volcanoes forming a mid-ocean ridge allows material to rise from the molten mantle. So the youngest

Fig 1.2. The rifting of a continent may lead to the creation of new oceanic crust such as that which now forms the floor of the Red Sea, created by the splitting of the African continental block.

Fig 1.3. The collision of an oceanic plate with a continental plate leads to its subduction which generates eventual remelting deep in the crust, the creation of a volcanic arc and finally a mountain range.

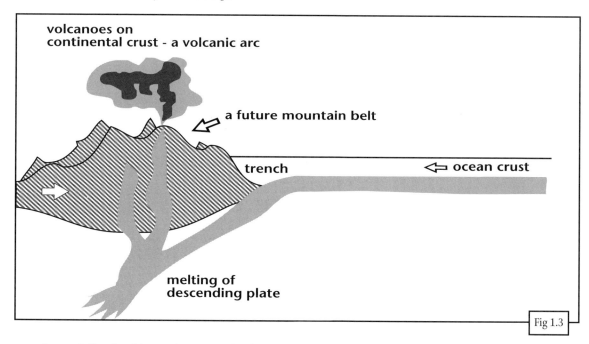

oceanic crust is found at this spreading centre, the oldest at the outer margins of the sea or ocean.

In contrast to this spreading process, plates sometimes move towards each other. They meet along a line called a collision zone. If a continental plate meets an oceanic plate a process called subduction occurs whereby the heavier oceanic plate sinks beneath the lighter continental one. These zones of collision, or subduction zones, are places of extreme stress which is why most major earthquakes are generated along them. What happens is that the build-up of strain is released by sudden movements of the crust which generate the friction responsible for earthquakes.

As the oceanic plate descends towards the mantle it becomes progressively hotter. Eventually it reaches a temperature where it begins to melt. As this solid plate becomes liquid, the lighter, molten material then rises and forms the source of volcanic eruptions on the continent, as shown in Figure 1.3. So both earthquakes and volcanoes tend to be concentrated along such zones of plate collision. An example of this process is the collision of the western edge of the American continent with the Pacific oceanic plate. Here, the two plates are sliding and scraping sideways past each other along the San Andreas fault resulting in the earthquakes which are a major problem for San Francisco and Los Angeles. The explosion of Mount St. Helens was also a result of this plate collision. On 18 May 1980, the height of this mountain was reduced by almost half a kilometre in one enormous explosion and a 400 square kilometre forest lay flattened around it. The people on the western seaboard of South America also suffer regular disasters as a result of the collision of these particular plates. The collisions around the margins of the entire Pacific plate and the ensuing volcanic activity has given rise to the term 'Ring of Fire' to describe this great circle of volcanoes.

When subduction has continued for some time, granites may be intruded upwards from deep within the continental crust as masses known as plutons. The rocks of the crust in these zones are squeezed and deformed and so the whole mass becomes thicker. The base of the mass is forced downwards, which results in the partial melting of continental crust to form the granites. The top of the mass is forced upwards, and it is this uplift which creates some of the mountain belts on the earth's surface.

Sometimes however, two continental plates may collide as in Figure 1.4. Such a collision means that neither is easily dragged downwards beneath the other since they are both of the same density. There is therefore considerable uplift of the resulting mountain belt. The Himalayas, for example, were formed relatively recently in the geological timescale by the collision of two such plates. Some 45 million years ago the Indian plate, moving northward, collided with the Eurasian plate. This generated the huge mountain belt which contains the highest mountains on Earth today.

Plate Tectonics therefore explains how the crust of the earth behaves as a continuous conveyor system. New crust is created at spreading centres and old crust is destroyed in collision zones. This process is thought to have been going on from an early stage in the history of the earth.

Classes of Rock

The crust of the earth is made up of a wide variety of rock types, but there are only three main classes of rock: igneous, sedimentary and metamorphic. Igneous rocks are those formed directly from the cooling of molten magma. These include the granites forced into the crust from below and basalts in cooled lava flows representing magma which has been forced out onto the surface. Those igneous rocks which cool in the earth's crust crystallize slowly and contain large, well-formed crystals. Those which cool on the surface crystallize much faster and thus will only contain very small crystals.

How is it then that we are actually able to see granites and other rocks which were originally at some considerable depth within the earth's crust? As parts of the crust are uplifted to form mountain belts they start to be attacked by the processes of erosion. The action of ice and water begins to break up the rock mass converting it to gravel, sand and mud. This process then eventually exposes deeper rocks on the surface.

Glaciers or rivers transport sediments from these uplifted areas until eventually they become deposited at the edges of the oceans. Thus the mountains

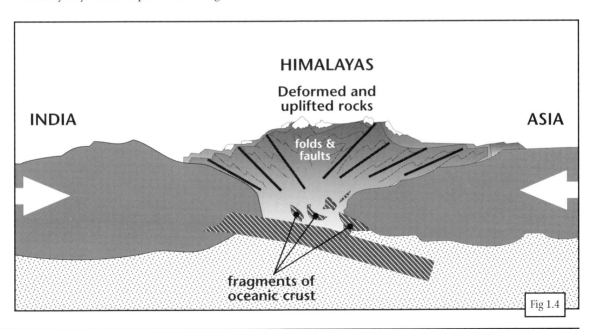

Fig 1.4. An ocean may be completely destroyed as two continents approach one another. The continents themselves will then finally collide and because of their low density neither will be subducted beneath the other. This will create vast mountain ranges. An example is the initial collision of India and Asia some 45 million years ago, eventually generating the Himalayas.

are worn down and their erosional products fill the adjacent sedimentary basins. If these sediments become buried under later sediment to a deep enough level in the crust they become converted, under increasing pressure, to sedimentary rocks. Sediments are laid down in layers. Each layer is called a bed and is the result of deposition under a particular set of conditions. Beds are separated from each other by bedding planes, which represent a change in the conditions of deposition.

Sir James Hutton, one of the founding fathers of modern geology, formulated a law in the late eighteenth century which had the rather Victorian-sounding title of the Principle of Uniformitarianism. This stated that essentially the same natural processes seen today have been active on the earth since earliest times. So to find out how sedimentary rocks were originally deposited we have only to look at modern environments and how they allow the transport and deposition of sediment in order to understand how processes operated in the past.

Both igneous and sedimentary rocks can eventually become a part of the crust which is subjected to exceptional stress and heating, in a collision zone for example. If this happens these rocks can be recrystallized and altered to become metamorphic rocks.

Geological Time

One aspect of geology which is important is the dimension of time. The earth is thought to have formed some 4500 million years ago. This we can call deep time. The oldest rocks so far discovered are approximately 4000 million years old. Whilst a rate of plate movement of 2 centimetres a year might seem slow, if one considers this process acting over a period of say 100 million years one can see that a plate may have moved 2000 kilometres in that time - a distance over half the width of the present Atlantic Ocean.

'As old as the hills' is an expression used to imply a great antiquity. But how do we know how old the hills really are? There are basically two ways to ascertain the age of rocks. The first is through the process of radiometric dating. Most rocks contain radioactive elements which are partly responsible for the background radiation that all living things experience on the surface of the planet. Unstable radioactive elements decay to form more stable 'daughter' elements. The time this process takes is known for these elements. So if we can measure the amount of undecayed material in a rock compared to that of the decayed material, we can arrive at an estimate of the age of formation of that rock. This method works best with igneous rocks whose radioactive clock starts ticking as soon as they solidify from the magma.

The second method of measuring the age of rocks is based on the fossils preserved in them. Throughout time, life on the planet has evolved. New species appear and existing species become extinct. The sedimentary rocks formed at a particular time contain a fossilized sample of the animals and plants living at that time. For example, if we excavated sedimentary rocks formed by deposition in a river valley some 100 million years ago, we might expect to find dinosaur bones. The same species of dinosaur would not occur in older rocks because they had not yet evolved. Dinosaurs became extinct about 65 million years ago. The fact that many fossils are time-specific can be used to match, or correlate, rocks from around the world which contain them. Nicolas Steno, a seventeenth century anatomist from Denmark, devised the Principle of Superposition. He reasoned that, in an undisturbed sequence, the sedimentary rocks at the bottom are the oldest and those at the top are the youngest. In geology many different groups of fossilized animals and plants are used for correlation of these rocks. Evolving species of calcareous algae or stromatolites are used in correlating and dating Precambrian rocks while a variety of marine shellfish are useful in Cambrian and younger rocks. Even fossil pigs have been used for correlation in the Pleistocene sequences of the East African Rift where they occur with fossil humans.

The study of the order of formation of sedimentary rocks is known as stratigraphy. In everyday life, time is divided into convenient, usable and comprehensible units from seconds to years. To discuss the huge expanse of deep time with equal convenience we need much larger units. The most commonly used interval is the geological period. Each period, like any measurement, must be defined with reference to an agreed yardstick or standard. The only standards available are the rocks themselves, so a period may be defined to encompass a sequence of rocks containing certain fossils. This sequence of rocks becomes a system. The definition of any period can be made by the first scientist who describes the rocks and their fossils. The geological periods vary in length but are usually in the order of tens of millions of years. These periods were assembled together, piecemeal, to encompass the whole of geological time. As new systems were described and placed in chronological order a more complete picture of the immensity of geological time with its diversity of biological evolution emerged.

The geological systems which form the basis for deep time are shown in Figure 1.5. The scale shows the time at which each period begins and some important events which occurred within it. Notice that the periods are not divided evenly. For example the Precambrian occupies more time than the rest of the periods put together.

Over these almost unimaginable periods of time the continents and oceans have continued to move, to be created and destroyed. The configuration of continents we see on the earth today is just a snapshot, a moment in geological time. To understand the geological history of Ireland, therefore we must understand these plate movements. Parts of the crust we see exposed today in Ireland may have originated far away from their present sites. This country carries in its rocks the evidence of the creation of oceans, the collision of continents and the other cataclysmic events that have shaped its beauty.

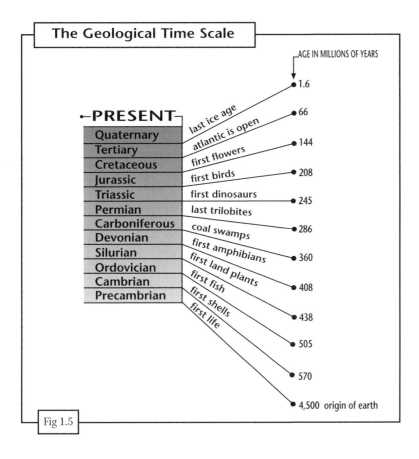

Fig 1.5. Some of the principal events on the geological timescale.

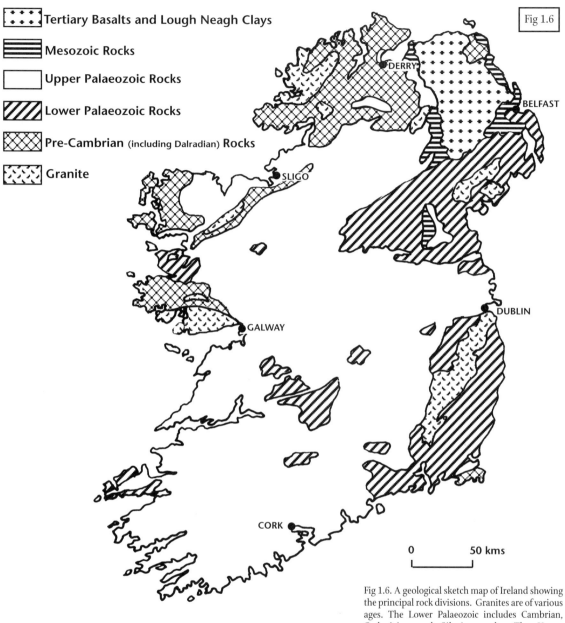

Fig 1.6. A geological sketch map of Ireland showing the principal rock divisions. Granites are of various ages. The Lower Palaeozoic includes Cambrian, Ordovician and Silurian rocks. The Upper Palaeozoic includes Carboniferous and Devonian rocks. The Mesozoic on this map includes the Permian, Triassic, Jurassic and Cretaceous rocks for simplicity; although strictly speaking the Permian is stratigraphically in the Upper Palaeozoic.

A more detailed geological map of Ireland is produced by the Ordnance Survey, Phoenix Park, Dublin. Geological maps of various parts of Ireland are available from the Geological Survey of Ireland, Haddington Road, Beggars Bush, Dublin, and from the Geological Survey of Northern Ireland, 20 College Gardens, Belfast BT9 6BS.

In the Beginning

The Oldest Irish Rocks

The Origins of Connemara

Life in the Precambrian

CHAPTER TWO

In the Beginning

*In geological history, as in the history of most human empires,
it is difficult to point out any definite commencement.*
J.B.Jukes

As we peer back through the mists of geological time, our perceptions become dimmer, our database more scanty. Very old rocks, like those of the Precambrian, conceal the clues to their origins more effectively than younger rocks. Because of the nature of the evolutionary process, fossil diversity and quantity increases with time. In the Precambrian rocks, fossils are rare and usually microscopic in size. This is not to say that animal life was not well established thousands of millions of years ago, only that most early life-forms were soft-bodied with no mineralized

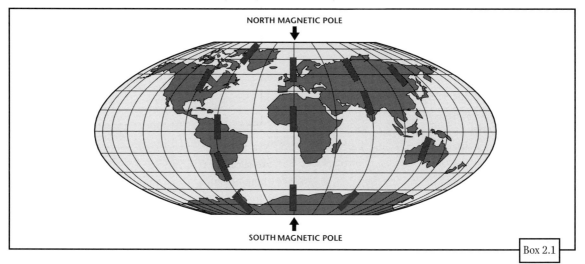

Box 2.1

Box 2.1 Palaeomagnetism

As well as internal structures and fossilized life forms, a rock may also contain "fossilized" magnetism. When a rock is forming, whether it be a sediment being deposited, or magma crystallizing, the iron minerals within the material become aligned along the magnetic field of the earth. The diagram shows the orientation of bar magnets at various points on the earth today. As the sediment or magma hardens to rock, the iron minerals are locked in place and point to the north (or south) magnetic pole of the earth at the time the rocks are forming. Thus if we examine the orientation of iron minerals in rocks say 100 million years old, we might find that they are aligned towards some point which is different from the north magnetic pole today. So either the magnetic poles of the earth have moved in the past 100 million years or the rock itself has moved from its original position. If however we were to examine several samples of rocks from different parts of the earth, each 100 million years old, we would find that each indicated a different pole position. Clearly it is not possible for the earth to have more than one north magnetic pole at any one time. Therefore, we must conclude that it is the rocks themselves which have moved from their original positions and not the magnetic poles. This confirms the process of continental drift. We can therefore use palaeomagnetism to deduce the relative positions of continents in the past. The data from such studies complements that derived from the past distributions of plants and animals that first suggested the movements of continents through time. The results can be used to construct palaeogeographic maps; that is maps which show the configuration of the earth at any point in past time. Unfortunately the further back we go, the less reliable these methods become.

skeletal structures. The chances of such delicate, and often minute, organisms being preserved are low. Added to this, these ancient rocks have often been through several episodes of mountain building. They are therefore very deformed, they have frequently been metamorphosed more than once and their structure is complicated and difficult to understand. Finally, these old rocks have been buried deeply by the younger, overlying, rocks. The chances of these older rocks being exposed on the surface of the earth are therefore much less than those of younger rocks.

The Oldest Irish Rocks

Fortunately there are places in Ireland where such ancient rocks are exposed. These rocks are not dated by fossils but by radiometric means. One such place is the small island of Inishtrahull off the north Donegal coast. The rocks exposed here are called gneisses, (pronounced with a silent 'g'). A gneiss is a rock that has been so severely metamorphosed that the original material has separated and recrystallized into

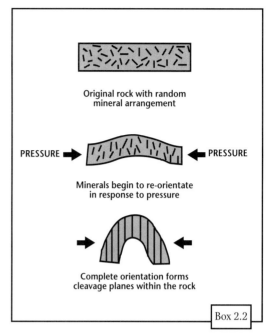

Box 2.2

distinct layers. Each layer may therefore be characterized by a particular mineral assemblage. There may, for example, be layers which are quartz rich alternating with layers rich in the mineral mica, a group of shiny minerals which form in thin, translucent, plates. Because these gneisses are often rich in quartz and have been metamorphosed and deformed at very deep levels in the crust, they are thought usually to represent pieces of continental, rather than oceanic, crust. The age of these rocks on Inishtrahull is about 1700 million years old.

In northern Mayo there are exposures of a group of rocks known as the Erris Complex. Whenever an association of rocks is termed a 'Complex' by

Box 2.2 Metamorphism

Sometimes rocks are modified by the process known as metamorphism. All Dalradian rocks, for example, have been through this process. Metamorphism occurs when the temperature or pressure or both, become so high that the rock recrystallizes from its original state, creating a new rock type. For example, a quartz-rich sandstone will become a quartzite; a limestone will become a marble; a mudstone will become a pelite. It is a gradational process so that at higher levels of metamorphism rocks may be converted to schists, or higher still to gneisses. At all times, however, the rock behaves as a solid, albeit sometimes plastically. If it melts completely, it then becomes part of the igneous process. Rocks may be metamorphosed by being buried deep within the crust, or by simple contact with a hot igneous body at any level in the crust.

The intense pressures often encountered in metamorphism cause the minerals in the rock to become aligned in response to this pressure. Eventually the minerals are all aligned within the rock and the rock tends to split along the planes defined by the mineral orientation. An example of this is slate, a low-grade metamorphic rock which splits along its cleavage planes. If the heat is sufficient to cause new minerals to grow along these planes the rock becomes a schist. This cleavage or schistosity is known as the fabric of the rock. Metamorphic rocks, like those of the Dalradian, are usually intensely folded and carry one or more fabrics.

Fig 2.1. The ancient supercontinent of Pangaea as it was in the Precambrian some 1000 million years ago. This is a palaeogeographic map.

geologists, one can be fairly sure that the relationships between the various rocks in the complex are not fully understood! The Erris Complex contains gneisses of two different ages: some are 1900 million years old and others are about 1300 million years old.

The oldest rocks so far dated in Ireland are found just south of Rosslare in County Wexford near the south eastern coast of the country. These rocks, part of the Rosslare Complex, contain gneisses dated at around 2400 million years. The Ox Mountains also contain some examples of these very old rocks.

The metamorphism of rocks results in the growth of new minerals which are stable under certain conditions of temperature and pressure. The radiometric dating of such metamorphic rocks usually gives us the date of the last mineral growth in that rock, not the date at which the sediment or magma became a rock in the first place. It is the date of the last metamorphic event, so the actual formation of the rocks which became these gneisses took place probably well before the date of metamorphism.

These fragments of continental rock form the very foundations of Ireland. We could call them the basement rocks. However it might be a mistake to assume that these fragments have always been in their present position. It is believed that at some time during the Precambrian these ancient mountain belts became amalgamated to form a single supercontinent called Pangaea. This large continental block would have included those pieces of the crust now comprising North America, South America, Africa, Antarctica, Australia, Europe, China and Siberia. We do not

Photo 2.1. Slices of rock may be ground down thin enough to become transparent. When viewed under a geological microscope, which uses polarized light, different minerals exhibit different optical properties and so can be identified. This is a thin section of a schist, a metamorphic banded rock typical of the Irish Dalradian. The lighter coloured minerals are micas which have grown and orientated themselves in response to heat and pressure. The width of this slide is about 3 millimetres.

know whether the processes of subduction and mountain building were exactly the same during early earth history as they are today; the crust was far less developed then and may have behaved differently.

The Origins of Connemara

Younger rocks than these basement fragments, although still of Precambrian age, form some of the best known scenery in western Ireland. The Twelve Pins of Connemara, Achill Island in Mayo and Errigal Mountain in Donegal consist of Dalradian rocks. The term Dalradian does not signify a stratigraphic system or period. Rather it denotes an association of rocks which have undergone the same sedimentary and metamorphic history. These rocks make up the Dalradian Supergroup, named after the ancient kingdom of Dalradia in Scotland, where they were first described. Their age is not particularly well defined but they were probably deposited and deformed initially before 590 million years ago. These Dalradian rocks constitute large parts of Connemara, northern Mayo, Donegal, north-eastern Antrim and Scotland. They are not exposed south of Galway Bay in Ireland. They are not as severely deformed and metamorphosed as the older gneisses and they still preserve some of the clues to their origins.

Photo 2.2

To describe Dalradian sedimentation it is convenient to divide the rocks stratigraphically into three parts: the Lower, Middle and Upper Dalradian. In total this Supergroup is over 14 kilometres thick.

The Lower Dalradian is dominated by quartzites. These were originally quartz-rich sands which were well sorted. This means that the sand grains are roughly the same size. To create such a well-sorted sediment, the sands must be continually reworked by currents over a long period of time. This reworking sorts the sediment by winnowing away the smaller grains and leaving the coarser ones behind. One environment in which this process operates efficiently is a shallow continental shelf. Here waves, tides and storms affect the sediment day after day, year after year. A lengthy process of sorting breaks down the weaker minerals and leaves the stronger, more resistant ones. Quartz is a very stable mineral; it is not readily broken down by either chemical or mechanical attack. Thus sediments which are well sorted and quartz-rich are called 'mature'. These mature sands are typical of sediment on many modern continental shelves.

Photo 2.2. A slab of Connemara Marble. The original limestone, of a type known as dolomite, was metamorphosed. This resulted in the growth of new minerals in the rock, such as serpentine which gives a green colour.

> **Box 2.3 Formation of Sediment Ripples**
>
> Currents move sediment, especially sand, through the formation of sand ripples on the substrate. The sand rolls up one side of the ripple until it reaches the top where it may accumulate for a short period. Eventually it avalanches down the steep side of the ripple, forming a thin layer, or lamination, of sand. This process is repeated for the lifetime of the ripple which loses sand from its upstream end and accumulates it on its downstream side. Thus it migrates in the direction of the current. When this sand layer is buried and becomes a sandstone it may preserve these laminations inside the bed. These are known as cross-laminations.
>
> In three dimensions these laminations are planes. They dip downwards in the direction in which the current flowed and if the dip direction can be measured, we can determine the direction of the ancient current which formed the ripples in the first place. These currents are known as palaeocurrents. Dalradian quartzites frequently show palaeocurrents from a variety of directions, a type of pattern which is typical of shallow marine sediments which are affected by tidal currents in opposite directions, storm currents in another direction and possibly longshore currents in another. The ripples themselves may also be preserved on the tops of beds of sedimentary rocks.

The Dalradian quartzites also exhibit cross laminations with variable palaeocurrent directions, shown in Box 2.3. Again this is typical of shallow marine sediments. So we might infer that the Lower Dalradian sediments were deposited on an ancient continental shelf and subject to the same sorts of tidal processes which operate today off the coast of Ireland. These quartzites are very resistant to weathering and so frequently form high conical mountains like the Twelve Pins of Connemara.

Towards the end of the deposition of the Lower Dalradian, conditions on this shelf changed. The amount of sand and mud being fed onto the shelf diminished dramatically. This allowed the deposition of lime-rich muds which became limestones. Some of these limestones were metamorphosed by heat and pressure to form the well-known Connemara Marble. Marbles of this age are found from Connemara to Scotland.

At the base of the Middle Dalradian rocks, overlying the marbles, is a sequence known as the Boulder Beds. These are rocks which contain large boulders of older rocks. Such rocks are called conglomerates. These beds can be found from Connemara, through Donegal and across to Scotland. They are exposed at Cleggan Head in Connemara and at Fanad in Donegal amongst other places. Similar beds of the same age are found in Greenland, Spitzbergen and Norway although in those places they are not called Dalradian. These beds reflect a particular event which affected a large part of the earth in the Late Precambrian and still show some clues to their origin. First, the boulder beds themselves are very poorly sorted. They contain a wide variety of sediment grain sizes from boulders down to mud. They were therefore unlikely to have been deposited by currents in shallow water, which are usually good sorting agents. Secondly, in between the boulder beds are horizons which contain drop-stones. These are pebbles or boulders enclosed in a finely layered mud or silt. The layering under the drop-stone is buckled downwards whilst that above is unaffected. A mud cannot settle out of suspension from water if there are currents flowing which are capable of transporting boulders, so the boulder must have been deposited by some mechanism not involving currents. That mechanism is floating ice.

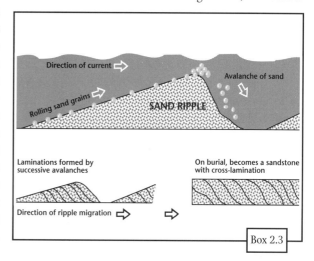

Box 2.3

The ice floats over the sea, or lake, bed which is slowly accumulating mud and silt. As it melts, the detritus it contains drops through the water column onto the mud and becomes a drop-stone. Later the stone is covered by subsequent layers of mud.

This and other evidence is taken to show that these boulder beds were deposited by glaciers or processes operating in a glacial environment. Glaciers do not sort sediment, they simply drag everything along with them. When they melt, they leave behind dumps of unsorted sediment as till; a rock formed from this material is a tillite. These Precambrian tillites show that a large part of the earth was subjected to a major glaciation about 680 million years ago. They are associated with rocks which were originally limestones which shows that the glaciers were thick and extensive enough to reach even warm latitudes, where limestones normally form.

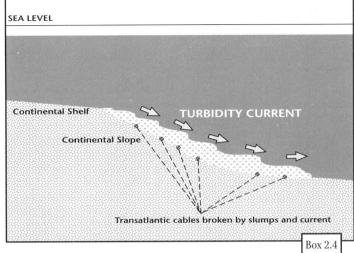

Box 2.4

After this major glacial episode, sedimentation on this Precambrian shelf temporarily returned to normal with the formation of quartzites and marbles. However, later, during deposition of the Middle Dalradian there is evidence that the shelf began to collapse. The shallow shelf sediments of the lower half of the Dalradian were overlain by sediments deposited in deeper water. Associated with these were volcanic eruptions possibly fed by magma escaping along faults where the shelf was cracking.

Box 2.4 Turbidites

Turbidity currents move down slopes because of their density which is higher than normal sea water due to the sediment they contain. They are not driven by water currents but by gravity. They may be generated by storms on the continental shelf, stirring up large quantities of sediment, or by earthquakes shaking the sediment on the continental slope itself. They were only discovered this century, and it was realized that they were one of the primary means of infilling deep-water sedimentary basins both today and in the geological past. Normal rates of sedimentation in deep-water environments are very low.

Of course they are very difficult to see in action. Their occurrence is unpredictable and their environment is often inhospitable to scientific observation. However, on 18 November 1929, a series of breaks occurred in transatlantic underwater telegraph cables off the coast of Newfoundland. The initial breaks were caused by an earthquake and all occurred at the same time near the continental shelf. Those cables in progressively deeper water, however, were broken after the earthquake and in succession. The last cable break was over 700 kilometres from the centre of the earthquake and 13 hours after it occurred. Since the time of each break was recorded, it was possible to calculate that the turbidity current that caused them moved down the slope at over 80 kilometres an hour and slowed down rapidly as it hit the less steep ocean floor.

The fossilized sediments from such currents are called turbidites.

The deep water sediments were deposited by two different processes. First in the top half of the Dalradian there are numerous pelitic rocks, fine grained rocks formed from sediment deposited out of the reach of waves or strong currents. Secondly there are metamorphosed sandstones formed by deposition from submarine turbidity currents. These are terrifying, catastrophic events, heavily laden with sediment and capable of moving down an incline, such as the continental slope, at speeds of over 100 kilometres an hour. Sandstones deposited by such a mechanism are called turbidites. They are important in that they fill the deeper parts of sedimentary basins where currents are generally very weak, and sedimentation is normally very slow. Sedimentation towards the top of the Dalradian took place in a variety of environments from deep to shallow water. This reflected the general instability of the shelf as parts of it collapsed faster than others.

Towards the end of the Precambrian, the whole of the Dalradian Supergroup suffered extensive deformation and metamorphism. The rock mass was distorted into huge folds, heated, and intruded by various igneous bodies derived from magma. The history of the Dalradian apparently records the typical phases of continental rifting perhaps with the opening of an ocean, followed by the initiation of subduction by collision and the formation of a mountain belt. Studies of palaeomagnetism and palaeontology suggest that the giant Precambrian supercontinent began to break up into different fragments towards the end of the Precambrian. For Ireland, two of these fragments are important: one was a continent called Laurentia and the other a giant continent called Gondwana. They began to separate from each other at this time and an ocean grew between them. This ocean was the precursor of the present Atlantic Ocean and is called Iapetus: because the Atlantic is named after Atlas, the giant of Greek legend, so geologists named this ancient ocean after his father.

Life in the Precambrian

Single-celled life-forms, without nuclei, probably existed on the planet from over 4000 million years ago. After 3,500 million years these organisms began to secrete calcium carbonate, the products of blue-green algae (cyanobacteria) or stromatolites. These are cabbage-shaped growths of algae, sometimes over a metre high. The modern descendants of these ancient life-forms were discovered living in Shark's Bay, Australia, only recently. Since they reproduce by simple cell division, parts of the first stromatolites are in theory alive today. These organisms lived by photosynthesis and by the release of oxygen into the atmosphere they prepared the way for the subsequent evolution of air-breathing animals. Many Precambrian limestones were formed by layers of carbonate deposited by cyanobacteria.

More sophisticated, but still minute, organisms were established by 2000 million years ago. These were made of clusters and strings of cells with nuclei, a major advance on their precursors. These fossils have been extracted from Precambrian rocks in many parts of the world including Australia, Canada and the USA and include various types of bacteria and fungi.

Much later, about 700 million years ago, the first true metazoans, organisms containing different tissue types and a variety of specialist organs may have appeared. Relatively large, soft-bodied, multicellular animals comprise the so called Ediacara fauna. These fossils, first identified in the Ediacara Hills of southern Australia and Namibia and since then found in all the continents, include a variety of animals mostly similar to jellyfish and seapens.

Very few Precambrian life-forms had hard mineralized skeletons. Skeletalized organisms did not appear until the start of the Cambrian, about 580 million years ago. Their advantages over their soft-bodied ancestors were substantial as shells provided support and protection from predators. Metazoan life was here to stay.

The Birth and Death of an Ocean

Ordovician Volcanoes

A Silurian Flooding

CHAPTER THREE

The Birth and Death of an Ocean

Tongues in trees, books in the running brooks, Sermons in stones, and good in everything.
William Shakespeare

So what was Ireland in the Precambrian? The earliest rock record shows that initially parts of what is now Ireland consisted of fragments of a supercontinent which had already suffered several mountain building events. The opening Iapetus Ocean began to be subducted at one of its margins deforming the Dalradian rocks and beginning a process that was to affect Ireland for the next 250 million years. Its evolution now became intertwined with the birth and eventual death throes of the major ocean, Iapetus. The Cambrian Period was typified by the gradual break-up of the supercontinent, Pangaea. Unfortunately there are few rocks of proven Cambrian age in Ireland. The Bray Group crops out discontinuously around the Dublin area, at Bray in County Wicklow and in a band running south-west through Enniscorthy near Wexford in the south-east of the country, but in western and northern Ireland there are no proven rocks of this age.

The continental fragmentation led to the creation of the Iapetus Ocean, which separated the northern and southern parts of Ireland by the beginning of the Ordovician Period, around 500 million years ago. It is difficult to be precise about the width of the ocean at its maximum extent but in all probability it was between 4000 and 6000 kilometres wide. The original evidence for the presence of the ocean came from palaeontology.

In the mid 1960s, a Canadian geophysicist, Tuzo Wilson, made a radical proposal that the present Atlantic Ocean was only the second of two major seaways that had separated the continents of Europe and North America during the last 1000 million years. He recognised a problem in that the fossils of Ordovician age occurring in North American rocks were different from those in Scandinavia and most of Europe. In particular the brachiopod, trilobite and graptolite faunas of both areas originated in two different faunal provinces, separated by what he called the proto-Atlantic Ocean, now called Iapetus. In modern oceans most faunas show a marked provinciality, or geographical distribution, controlled by climate; polar faunas are quite different from those of the tropics. Also animals living on the continental shelves are more restricted in their distribution than those of the ocean depths which are more cosmopolitan. During the early Ordovician the less mobile groups such as the brachiopod shellfish and the trilobite arthropods were more sensitive to climatic controls than say the graptolites which floated with the plankton in the surface waters of this ancient ocean or moved up and down the water column during bouts of feeding. Thus the more mobile groups like the graptolites had a better chance of crossing this wide barrier to migration than the members of the fixed benthos such as the brachiopods.

Ordovician Volcanoes

In the Ordovician Period both halves of Ireland became the sites for numerous and long-lived belts of volcanoes. In fact the geology of these rocks is often dominated by the products of different sorts of volcanoes. Along the northern margin of this ocean at this time there existed a sedimentary basin whose history is almost completely preserved in the geology of southern Mayo. This particular sequence of Ordovician rocks is about 12 kilometres thick, and forms the high plateau of the Partry and Sheeffry Hills framing the wild grandeur of the mountains around Doo Lough. The oldest rocks in this succession are exposed around Lough Nafooey near the southern border of County Mayo. These rocks, the Lough Nafooey Group, consist of dark green pillow lavas, with some black shales and coloured cherts. This is part of a classic association of rock types and often indicates the formation of such rocks on a deep ocean floor. The pillowed shape of these fossilized lava flows is formed when the hot magma escapes through cracks in the crust and is extruded into relatively cold water. The heat of the magma causes a jacket of steam to be formed around it at its contact with the water. The lava breaks up and separates into individual globules or pillows as it oozes onto the sea floor, each pillow surrounded by its own steam jacket. The black shales associated with these lavas were formed by

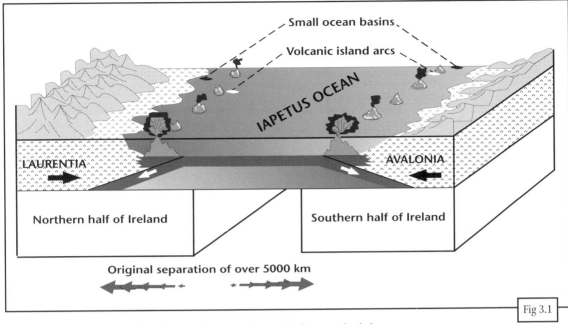

Fig 3.1

the very slow settlement of mud onto the ocean floor, which entombed the remains of graptolites. These remains can be dated and show the age of the volcanism to be Tremadoc, that is very early in the Ordovician.

The chemistry of these lavas indicates that they may have formed in an island-arc environment. A volcanic island-arc appears on a map as a crescent shaped line of volcanoes (a straight line drawn on a globe becomes a curve when projected onto a two-dimensional map). The arc is generated on oceanic crust above a subduction zone so the ocean crust of Iapetus had now begun a process of subduction as shown in Figure 3.1.

Later in the Ordovician these volcanic edifices were swamped by a massive input of sediments infilling the arc-related basin. At first these were deep-water sediments deposited by turbidity currents which formed an enormous thickness of conglomerates, sandstones and siltstones. Later, by the middle of the Ordovician, this part of Mayo was uplifted as the arc collided with the continent. This very mountainous area became a site of extensive erosion to form alluvial fans at the base of the mountains. Such fans are wedge-shaped masses of sediment formed by erosion and deposition of rapidly uplifting mountains. The basin had thus become infilled with marine sediment which was covered by the coarse gravels and sands of the fans deposited largely on land. Volcanic activity continued throughout the history of the basin. Interbedded with the conglomerates and sandstones of the fan deposits are thick tuffs or ignimbrites. These are rocks formed by the settling of the dust and ash from the explosion of volcanoes.

Fig 3.1. The approach of the continents of Laurentia and Avalonia resulted in the eventual destruction of the Iapetus Ocean and the creation of Ireland as a coherent part of a western European crustal block.

Photo 3.1. Pillow lavas representing the fossilized floor of the Iapetus Ocean. These are early Ordovician lavas and the photograph shows several pillows, one of which has the lens cap near its centre.

Photo 3.1

Fig 3.2. A view over southern Mayo in the early Ordovician. The waters of the Iapetus Ocean lap against a belt of active volcanoes.

These alluvial fans, which covered much of the northern half of Ireland, would have presented dramatic landscapes. Their surfaces of sand and gravel would have been completely devoid of any vegetation since land plants had not yet evolved and these great plains of outwash would have been ravaged by flash floods of enormous intensity. Great explosive clouds of incandescent ash and dust would have periodically swept over their surfaces, eventually settling to form glowing layers of debris up to 10 metres thick. These explosions would have been of the same type as the one which buried the town of Pompeii under ash and blocks of lava in 79AD.

Other Ordovician rocks, part of the northern margins of Iapetus, are found on the small islands of Lettermullen and Gorumna in western Connemara. At the base of this sequence are pillow lavas and thick horizons of coloured cherts. These are overlain by conglomerates and sandstones derived from a continental land mass. It is difficult to be specific about the origin of these particular rocks since they are quite deformed and are in fact upsidedown in a stratigraphic sense. They probably represent a suspect terrane with respect to the Ordovician rocks of southern Mayo. However these rocks have been likened to an area of

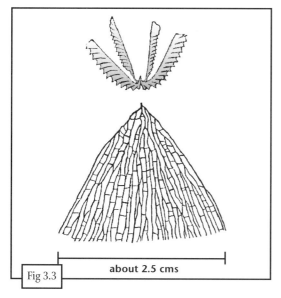

Fig 3.3. Two graptolites, small colonial organisms which were common in early Palaeozoic seas. They either floated in surface waters or moved up and down the water column during feeding bouts. Most graptolites are excellent for dating rocks since they evolved rapidly and were geographically widespread. The upper one is a graptoloid and the lower a more primitive dendroid graptolite.

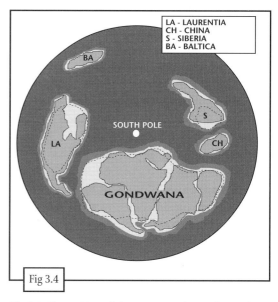

Fig 3.4. The position of the continents during the Cambrian Period, about 570 million years ago.

rock known as the Longford-Down Massif, a thick succession of Ordovician and Silurian rocks outcropping in a wide belt extending from Longford town towards the north-east coast around Bangor. These rocks, apart from early volcanics and cherts, are dominated by turbidites deposited in deep oceanic waters. Some geologists believe that these sediments were formed in what is known as an accretionary prism. Such prisms are thick successions of sediments deposited and finally trapped above a subduction zone. So parts of the old ocean floor, in the form of the pillow lavas and cherts, were scraped up into this prism along with turbidites derived from the adjacent continent.

In Clew Bay there are discontinuous outcrops of rocks which until recently were thought to be Cambrian or even Dalradian. Microfossils such as the remains of plant spores and tiny marine organisms extracted from these rocks have shown, however, that they are Ordovician and Silurian. These rocks are exposed in the north-western quarter of Clare Island and in a coastal strip on the southern shores of Clew Bay. They are collectively known as the Clew Bay Complex, and complex they are. It appears that at least some of them consist of a type known as melange. A melange is a rock which is made up of large blocks of different rock types set in a shaly matrix. Some melanges contain blocks the size of a village and the blocks are often arranged in a mixed up and random fashion.

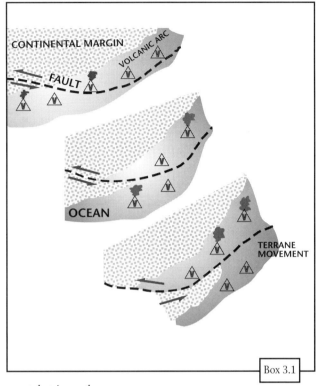

Box 3.1

Box 3.1 Suspect Terranes

During the collision of crustal plates the stresses involved often result in plate fragments being rotated or moved out of their original position. These movements take place along major fault zones which may be up to thousands of kilometres long. So fragments of crust may end up a long way from where they started. If we find two or more areas of rock which apparently bear no relationship to each other they may be suspected of being areas which have moved in relation to each other, and they therefore become suspect terranes. In the diagrams we see the effect of a single such movement along a fault zone. The result is to duplicate fragments of continental crust and the volcanic arc system. This kind of movement has been used by some geologists to explain the position of the Connemara Dalradian, for example. In the lowermost diagram the Connemara Dalradian could be the fragment of continental crust moved along a fault zone during the Ordovician. Thus the Connemara rocks have become separated from their 'parent' body, the Dalradian of Mayo, Donegal and Scotland. Today we see that these old Connemara rocks are separated from their equivalents in Mayo by the younger volcanic and sedimentary rocks of the southern Mayo Ordovician. Plate collisions almost invariably involve the movement of such terranes leading to quite complex reshuffling of continental and oceanic fragments.

The Clew Bay melange consists of blocks of chert, sandstone and volcanics in a black shale matrix. Melanges are usually generated by the sudden collapse of rocks or sediments caused by severe instability in the area. Such an unstable environment is to be found, for example, directly above a subduction zone where two plates are colliding. Many melanges are directly related to subduction zones. The newly formed strata on both the oceanic and the continental crust are disrupted by the collision. They collapse in large sheets or fragments into the black muds of the ocean floor. Here they may be further deformed as the collision proceeds. It may be that these rocks around Clew Bay represent a fossilized subduction zone, the actual contact between the north Iapetus plate and the Laurentian continent in the late Ordovician or Silurian.

Similar events were happening on the southern margin of the Iapetus Ocean which was defined by the continent known as Avalonia. Avalonia was a part of the larger continent of Gondwana which had broken away and drifted northwards towards Laurentia. Rocks of mid - late Ordovician age are present around the Wicklow Mountains with numerous examples of volcanic rocks preserved there. Lambay Island off Dublin is also a remnant of this volcanism as is the Chair of Kildare, which is capped by limestones. So in the Ordovician both edges of the ocean were being subducted beneath the two facing continental blocks and consequently the width of the ocean was progressively decreasing throughout the period. The volcanic arcs eventually collided with the continents and the Ordovician rocks were deformed before the Silurian sea began to flood over their partly eroded

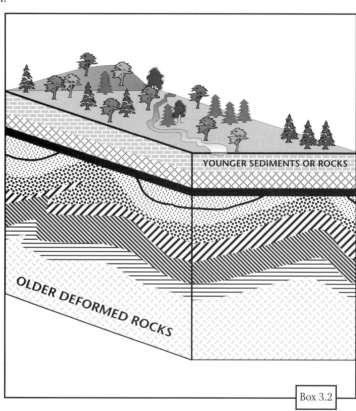
Box 3.2

Box 3.2 Unconformities

An unconformity represents a missing piece of earth history. Once rocks have been buried and deformed in the earth's crust they may eventually be uplifted to reach the surface of the earth again. They may then be overlain by younger deposits which themselves become rocks upon further burial. Thus the plane that separates these older and younger rocks represents a time interval, which must be long enough to deposit the older sequence, bury it, deform it, uplift it again and partly erode it, before the deposition of the younger layers. So unconformities can represent tens of millions of years of the geological record which have been removed. One of the first unconformities to be recognized was that at Siccar Point in Berwickshire, Scotland. Here Sir James Hutton recognized steeply dipping Silurian rocks, 'the ruins of an earlier world', overlain by nearly horizontal Devonian rocks. He illustrated this in his *Theory of the Earth*, published in 1788.

remnants. The relationship between the Ordovician and Silurian rocks in Ireland is therefore often an unconformity.

A Silurian Flooding

By the beginning of the Silurian Period the Iapetus Ocean still existed but was narrowing rapidly by continued subduction. Throughout geological history the battle between land and sea has resulted in constant changes in relative sea levels. As an example of such changes, we can see, in the Silurian rocks exposed from the Atlantic coast eastwards to Lough Mask in northern County Galway, the results of a classic transgression, or marine flooding, over eroded Ordovician and Dalradian rocks. Being one of the best exposed fossilized transgressive sequences in Ireland, the record of events is preserved in some detail. Clues preserved in these rocks allow one to decipher the nature of this marine flooding. The stratigraphic succession here is shown in Figure 3.5.

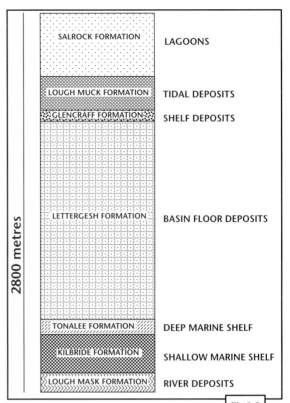

Fig 3.5

The oldest Silurian rocks, the Lough Mask Formation, are bright red sandstones with cross-lamination well developed. This formation does not contain fossils and is interpreted as being the result of deposition in a large river system with high-velocity currents. Palaeocurrents demonstrate that these high-energy rivers flowed from the north at this time. The overlying Kilbride Formation begins with similar sandstones, but these contain vertical trace fossils. These are the preserved burrows of an animal, probably a type of worm. Since land-living animals were still very rare, this is taken to indicate that the river sands were occasionally covered by sea water bringing in the colonizing fauna to burrow for food. This was the very beginning of the marine transgression. Had we been standing there at that time we would probably have been at the margins of an estuary or delta made up of thick river sands which were covered during storms or exceptional high tides by the sea. The edge of this sedimentary basin was subsiding, however, and gradually the marine environment became permanently established here with the result that the overlying sandstones and siltstones of the Kilbride Formation contain a remarkable fossil fauna of marine organisms.

As the transgression proceeded and the sea deepened, the animal communities and their behavioural patterns changed as did the enclosing sediments. The oldest animal communities were dominated by a variety of gastropods (sea snails), some bivalves and the burrowing brachiopod *Lingula*. These intertidal communities were gradually replaced by the brachiopod *Eocoelia*, living in shallow-water, subtidal environments. These animals were low-level filter feeders relying on nutrients in suspension stirred

Fig 3.5. The Silurian succession of northern Galway which is the result of the flooding of a continental margin by the sea and the subsequent infill of the edge of the marine basin. The opposite of a transgression such as this is a regression, in which the sea level progressively falls in relation to the land surface.

up by turbulent near-shore waters. Higher in the formation, and therefore younger, the mid-shelf communities were dominated by corals. As the sea deepened further the corals were replaced by underwater forests of crinoids (sea lilies) with some trilobites and brachiopods. Finally, at the summit of the transgression the red siltstones of the Tonalee Formation preserve an impoverished fauna of a few minute brachiopods. This marginal fauna survived in relatively deep-water, dark conditions.

So by the end of the deposition of the Tonalee Formation water depths had increased to perhaps 100 or 200 metres. This gradual and gentle subsidence of the basin floor was now interrupted by faulting, which caused parts of the sediment layers to collapse. Large broken fragments of this shelf material are found in the conglomerates of the overlying Lettergesh Formation. These conglomerates grade upwards into a 1500 metre thick sequence of sandstones and tuffs deposited by turbidity currents. These currents flowed from the north into this deepening basin. The sediment itself was derived from a volcanic arc. This arc, unlike the earlier extinct Ordovician arc, was founded on continental rather than oceanic crust. It represents the last explosive effects of subduction in this area. The last vestiges of the mighty Iapetus ocean were now being consumed beneath both continents as they approached each other. The last three formations in this Silurian sequence represent deposition in progressively shallower water as the basin became filled with sediment.

A similar history can be seen in small outcrops of Silurian strata around Charlestown. Indeed this pattern of sedimentation can be recognized in rocks exposed in the Midland Valley of Scotland which may have formed part of the same Silurian basin.

How can geologists determine the depths of water present millions of years ago? Various structures in sedimentary rocks, such as ripple marks and cross-lamination, may help to determine the water depth and current velocity under which they were formed. The type of fossil present can also be of use in determining past environments. Communities of bottom-living marine organisms are known to inhabit certain definite water depths. As well as depth, other factors such as the type of sediment and the degree of turbulence can govern the distribution of these so-called benthic communities. An American palaeontologist, Alfred Ziegler, established a series of Silurian marine communities developed against a spectrum of various water depths. Each community was named according to the most common or distinctive member of the assemblage. More recently the term benthic assemblage zone has been employed to distinguish communities which inhabited various marine environments. These zones range from BA1 to BA5 and represent water depths from zero (shorelines) to about 100 metres (the outer continental shelf). In deeper water the community is referred to as BA6.

It is not just the benthic communities which change with increasing water depth. The behaviour pattern of organisms changes too. There are preserved marks called trace fossils in rocks which record the burrowing or

Figure 3.6 A typical Silurian sea floor scene.

moving of organisms through the original sediment. Very rarely can the makers of these marks be identified. Nevertheless we usually have a good idea of what the animal was doing. In stressful environments, such as the intertidal one where tides come in and out, waves break and storms whip up the sediment, many animals live in deep vertical burrows so that they can escape quickly from rapid changes in their environment. In subtidal environments exposure to the air is less likely and stress is reduced. So burrows are shallower and often inclined. Out on the deep shelf where conditions are more or less constant there is little need to burrow and the traces are dominated by horizontal feeding trails. Adolf Seilacher developed a type of 'community pattern' to these trace fossils which he related to water depth. These associations of trace fossils together with the preserved benthic body fossils allow us to determine marine water depths in fossiliferous rocks with a reasonable degree of accuracy.

Photo 3.2 Fossilized brachiopod shell moulds *(Eocoelia)* which inhabited shallow-water marine niches in the Silurian Period.

There are other Silurian successions in the Mayo region. For example the mystic mountain of Croagh Patrick consists largely of Silurian rocks. These are sedimentary and volcanic rocks. They include quartz-rich sandstones which form the peak of the mountain itself. The relationship of the rocks to the northern Galway Silurian rocks is not yet established, indeed they may not be in their original positions with respect to one another. Adjacent to the Croagh Patrick Silurian rocks are other exposed Silurian strata around Louisburgh and on Clare Island. These are shallow-water sedimentary rocks which may be younger than both the Croagh Patrick and northern Galway successions.

On the southern side of the closing ocean, the Dingle peninsula also preserves Silurian rocks. They are exposed on the dramatic cliffs on the western edge of the peninsula and on Inishvickillane, the most westerly of the Blasket Islands. These rocks are largely shallow marine in origin but also contain a thick volcanic sequence, well exposed on Clogher Head. As well as ignimbrites there are lava flows preserved as rhyolites (volcanic rocks containing large amounts of quartz). Many of the sedimentary rocks here are fossiliferous and this succession may well be the remnant of a number of volcanic islands situated in the Iapetus ocean in the Silurian.

Photo 3.3 Silurian fossils including brachiopods, corals and the long columns of crinoid stems (sea lilies).

Silurian rocks are also exposed around hilly areas in the central Irish plain, notably at Slieve Aughty, Slieve Bloom and the Arra and Galty mountains. These rocks are the preserved remnants of the infill of a large basin between the advancing continents in the Silurian. They are largely turbidites and were derived from both the Laurentian and Avalonian continental margins.

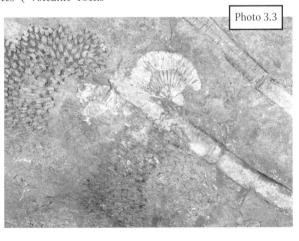

Photo 3.4 A microscope photograph (photomicrograph) of a Silurian tuff. The crystals of feldspar (light coloured) were blown out of the depths of an ancient volcano situated near Dingle in the Silurian Period. The width of this slide is about 3 millimetres.

Photo 3.4

The Silurian Period then, was one in which the Iapetus Ocean was finally consumed beneath the closing vice of two continental masses, Laurentia to the north and Avalonia to the south. The closing of the ocean can be traced by the changing provinciality of the fossils. Two British palaeontologists, Robin Cocks and Stuart McKerrow, published a score chart in 1976 plotting animal mobilities against the width of the ocean. During the Cambrian, only the most mobile of the animals, the dendroid graptolites, were similar on both sides of the ocean. During the Ordovician, as the sea began to close, the types of trilobite and brachiopod became progressively more similar. During the Silurian the seaway was considerably narrowed and easily crossed by most marine organisms. The final closure in the Devonian ended any provincial differences between the American and European faunas.

The Iapetus Ocean may well have contained islands like the Pacific Ocean today. Some rocks of this age contain exotic assemblages of fossils which cannot be directly related to fossils of the old platforms around America and Europe. These assemblages have so called endemic animals, not known from elsewhere. So they may have developed around islands in Iapetus similar to the Galapagos Islands of today, acting as stepping stones for the migration of some animals but also as sites where new species may have evolved.

The effect of this closure was the deformation of all rocks and sediments on both continental margins and in between the two continents. This deformation was the final phase of a long process which had begun in the Early Ordovician or before, called the Caledonian Orogeny or mountain-building episode. The final result was the creation of an enormous mountain range which extended from the east coast of North America, through Newfoundland to Ireland, Scotland and Scandinavia.

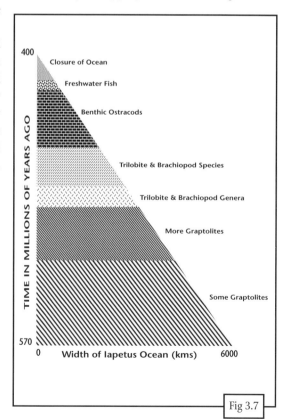

Figure 3.7 A chart showing the times at which various organisms became common on both sides of the Iapetus Ocean (after studies by Robin Cocks & Stuart McKerrow).

The Rise and Fall of the Mountains

The Great Caledonian Mountains

The Mountains Crumble

The Arrival of Life on Land

CHAPTER FOUR

The Rise and Fall of the Mountains

Then Abner Dean of Angel's raised a point of order, when
A chunk of Old Red Sandstone took him in the abdomen,
And he smiled a kind of sickly smile, and curled up on the floor
And the subsequent proceedings interested him no more.

Bret Harte

The Great Caledonian Mountains

During the Caledonian orogeny the rocks themselves were folded and faulted and finally uplifted so that by the end of the Silurian, Ireland would have been a mountainous region perhaps similar to the present-day Alps. This subduction and collision would have generated large amounts of heat in the crust. We can see the effects of this in the metamorphism of the rocks involved, and in the presence of large granite intrusions seen in various parts of Ireland. Granites are generated by the melting of continental crust and are therefore typical of the final phases of a mountain building episode where the shortening of the crust causes thickening. This thickening not only causes continental crust to rise at its upper levels but also to sink, and therefore heat up, in its lower levels. This melted material is less dense than solid crust and therefore rises, forming a granite intrusion. The largest intrusions of this age are the granites of Donegal, the Galway Granite in Connemara and the Leinster Granite south-west of Dublin. Granite often contains large crystals of quartz, feldspars and micas, their size being due to the slow cooling of these often huge bodies of magma.

Granites retain some of their heat for long periods of time and at some time in the future it is possible that they may be used to generate geothermal energy. Cold water may be pumped deep into a granite, become heated and then be abstracted for heating purposes. The costs would be minimal compared to other energy sources.

Collision zones in the earth's crust are often important sites of mineralization. Fluids are liberated at these zones as a result of the melting of both types of crust. These fluids move upwards to levels of lower temperatures and pressures. Here, as the fluids cool, the dissolved materials crystallize out of solution. An example of this process is the geological concentration of gold. Exploration for this mineral has taken place in many areas in Ireland, notably in Donegal, at Croagh Patrick and around Doo Lough in Mayo. Gold is present in a number of rocks but its concentration in quantity is very rare, hence its high value. It may be concentrated in a number of different ways. In the case of Mayo the gold is concentrated along shear zones cutting both Ordovician and Silurian rocks. The final concentration therefore probably took place late in the Silurian. Shear zones are areas where intense strain has occurred in the crust. They are often sites of fluid migration as deep-seated fluids under pressure are liberated. If a shear zone cuts deeply buried basic igneous rocks, the original gold contained by them migrates upwards with the fluids and is concentrated higher in the crust in areas of lower pressure. The Mayo gold is often found disseminated through veins of white quartz.

The line along which two continents finally collide in an orogeny is called the suture zone. The Caledonian suture zone in Ireland is thought to run from the Shannon estuary north-eastwards to a point about 30 kilometres north of Dublin. This zone then separates the ancient continents of Laurentia to the north and Avalonia to the south, with perhaps a few small island masses squeezed between them. It signifies the final geological amalgamation of Ireland into a unified crustal block, a block that was to remain virtually intact as a piece of continental crust for remaining geological time.

The Mountains Crumble

By the beginning of the Devonian Period, some 410 million years ago, the great Caledonian mountain belt, of which Ireland was a part, had begun to uplift rapidly in response to the collision of the two continents Laurentia and Avalonia.

The Himalayan belt is a recent orogen caused by a double continent collision. As a result the Himalayas are rising at rates of between 0.5 and 4 millimetres per year. If we assume that similar uplift rates applied in the geological past then over the 50 million years of the Devonian Period, the Caledonian belt may have risen by up to 20 kilometres. However, as mountains rise they become more rapidly eroded so we would not have expected to see mountains 20 kilometres high in the Devonian. The vertical movement of mountain belts is partly governed by a principle known as isostasy. Plates may be thought of as blocks of wood floating in water. The thicker the block the higher it rises above water level. If we place weights on a block it sinks lower in the water. When the weights are removed the block rises. Thickened crust is generated by plate collision creating a mountain belt. As the belt is uplifted the products of its erosion are dumped in adjacent basins. The floors of these basins subside to accommodate all this sediment. However as weight is removed from the upper levels of the mountain belt the base of the belt rises to compensate for this. So gradually as the orogen is uplifted it is worn down, eventually to its roots, and the basins become filled with new sediment which eventually forms new crust. Most of Ireland at this time consisted of an extremely mountainous landscape. As a result of the rapid erosion by wind and water these mountains rapidly shed their detritus onto alluvial fans which discharged into lakes or extensive river systems.

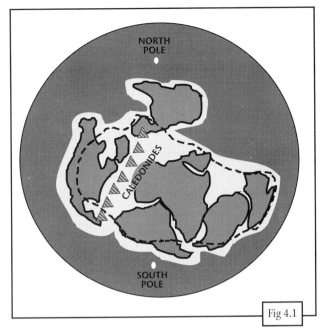

Because the underlying rocks had been involved in the Caledonian orogeny, Devonian strata may rest unconformably on older rocks. An example is found near Pomeroy where the Devonian is unconformable on Silurian strata. The best exposures of Devonian rocks in Ireland are found on the Dingle peninsula and farther south in a great east to west belt from the Atlantic coast of Kerry to the Irish Sea around Dungarvan. These rocks form the backbone of many of the mountains in this area, such as Mount Brandon in Dingle and McGillycuddy's Reeks around the lakes of Killarney.

Most Devonian rocks in Ireland consist of what is called the Old Red Sandstone. As the name suggests they are dominated by red coloured sandstones and conglomerates. Many of the sandstones show well developed cross lamination and some interbedded mudstones show fossilized shrinkage cracks, formed when newly deposited wet mud dried out in the air. Their presence shows that the sediments were deposited in environments which periodically dried out, that is, they were above sea level. Some of the sandstones may contain discontinuous layers of a white limestone known as caliche, or calcrete. These calcium-rich deposits are formed in the upper part

Fig 4.1. The positions of the continents during the Devonian some 400 million years ago. The Caledonian mountain belt formed by this time was already being eroded and the sediment was deposited as the Old Red Sandstone.

of the sediment profile by evaporation. If the water in the sediment contains dissolved calcium carbonate, this precipitates when the water is evaporated off. This feature also implies that the sediments were deposited above sea level, known as the subaerial environment, and that the Irish climate at the time was warm enough to induce regular and rapid phases of evaporation. In fact Ireland lay some 20 degrees south of the equator during much of the Devonian.

The sediments of the Old Red Sandstone were deposited on alluvial fans largely draining from the mountainous hinterland to the north and large parts of the country were buried beneath their red sands and gravels. Owing to the arid nature of the climate, punctuated by flash floods, large quantities of wind-blown sand often moved across the surface of the fans, choking nearby river systems. The sand was blown into large dune structures, migrating across the Irish landscape like those in the Sahara desert today. Since they migrate like giant ripples such dunes produce large scale cross-bedding. Fossilized examples of such structures can be seen for example on the Dingle peninsula today.

There are smaller outcrops of Devonian rocks in the northern half of Ireland. These are exposed just east and north of Westport, near Clew Bay, in a narrow belt running through Lough Gara at the borders of counties Sligo, Leitrim and Roscommon. These deposits may represent the infill of separate land-locked basins within the mountains of the Irish Caledonides.

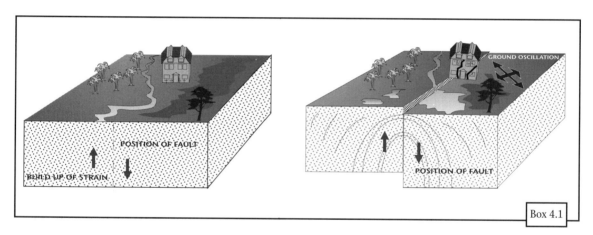

Box 4.1

Box 4.1 Earthquakes

Earthquakes are used by scientists to determine the internal structure of the planet. They are usually generated by movement along fault planes caused by a progressive build-up of strain in the rock mass which is released by sudden fracturing. This release generates shock waves (seismic waves) through the crust and may cause the surface of the earth to oscillate violently.

The strength of earthquake shock is often measured on the Richter Scale. This is an open-ended scale but natural earthquakes are unlikely to exceed a force of 12. However the scale is not arithmetic so that a force 2 quake is some ten times more violent than a force 1. Earthquakes are most common in areas of the world where plates are separating or colliding. The Los Angeles quake of January 1994 was an example of the oceanic Pacific plate moving past the continental American plate. There is little doubt that Ireland in the Devonian suffered intense earthquake shocks as a result of plate collision.

Earthquakes on the sea floor may generate large destructive waves in the water column, called tsunamis. They are sometimes called tidal waves although they are not generated by tidal forces. In deep oceanic waters they are hardly noticeable but on approaching shallow water they can achieve heights of many tens of metres and cause considerable devastation.

The Arrival of Life on Land

The creation of this Devonian supercontinent, straddling the equator, provided unrivalled opportunities for organisms to colonize warm landscapes for the first time. It is argued that forms of soils had already coated the land since the Cambrian, acting as ready-made seed beds but it was during the Silurian and Devonian that the main move from marine to terrestrial environments was made. In one of the most spectacular evolutionary events in the planet's history, plants, arthropods, molluscs and amphibians invaded the land. In Ireland, one of the first land plants was collected from Silurian rocks in the Slieve Bloom mountains. This primitive vascular plant, *Cooksonia*, is now known from many parts of the world in rocks of about the same age. These early plants had to develop a water circulatory system, a waxy coating to prevent water loss, and pores or stomata to allow respiration and transpiration. Additionally they had to develop structures to counteract gravity, having left the support offered in a water environment. Most of these innovations can be identified in *Cooksonia*. The rapid proliferation of this plant and its allies led to the covering of marine marginal areas with a green carpet of vegetation about 5 centimetres high. This greening of the land and the generation of plant litter made the subsequent invasion of animal life inevitable.

By the Late Devonian, 375 million years ago, many landscapes were covered with a lush green vegetation. Trees and the first forests had become established. Occasionally large lakes would have occupied areas near the toes of the alluvial fans or been associated with the large rivers. The deposits of one such system are fortunately preserved in Ireland in the form of the famous Kiltorcan Beds at Kiltorcan in County Kilkenny: a sequence of sandstones whose rich fossil flora has been known by geologists for over a century. This flora is key evidence of the rapid diversification of plants and trees during this time. It contains ten genera of plants, including the fern-like *Archaeopteris*, which was well over a metre high and records an early stage in the evolution of true tree leaves, *Cyclostigma*, an early lycopod or spore-bearing plant, and *Spermolithus*, one of the earliest seed plants. More recent research has suggested the Kiltorcan Beds also preserve a mosaic of swamp deposits. The swamps themselves were traversed by migrating streams carrying wood debris while lycopod trees flourished within the marshes leaving fossil root structures. *Archaeopteris*, possibly reaching heights of 20 m, and the arborescent lycopods, such as *Cyclostigma* growing to 5-6 m, thrived on adjacent, better drained, upland areas. There was no connection between the swamps and the developing ocean to the south.

These Kiltorcan Beds also contain the fossil fish *Coccosteus*, and the large freshwater mussel *Archanodon*, together with crustacean and eurypterid fragments. The Kiltorcan flora and fauna paint a vivid picture of life within and around an oasis in the generally more arid

Photo 4.1 A large fossilized sand dune of Devonian age. The beds of sandstone dip to the left. In the upper half, more steeply dipping planes represent the successive fronts formed on the leading edge of a migrating sand dune. Such planes are called foresets.

Photo 4.2 Branching stem of *Cooksonia*. This is one of the earliest land plants from the mid Silurian. Club-like spore pods are found at the end of the stems. Magnification about x10.
Photo Courtesy of Dianne Edwards

Photo 4.3 Part of a fossilized early lycopod tree from the Kiltorcan beds. The name of this fossil is *Cyclostigma* and it represents an early development of spore-bearing plants. The photograph is normal size.

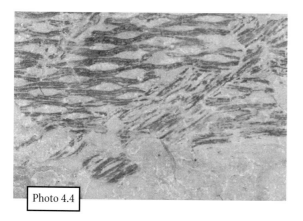

Photo 4.4 Fossilized bark of *Lepidodendropsis*. This was related to the giant lycopods of the Carboniferous Period. Magnification is x2.

Photo 4.5 The Inch Conglomerate from Dingle. A conglomerate is a rock containing pebbles and boulders of other rocks. The nature of these clasts can be used to determine where the conglomerate was derived from.

environment of the Old Red Sandstone times.

Elsewhere, on the northern coast of Valentia Island, County Kerry, there is further evidence of life on land. Over 150 footprints of an amphibious tetrapod are present in rocks originally deposited on alluvial plains. The trackway is sinuous and over 6 metres long. The spacing of the footprints suggest the animal was about one metre long and had a stride of about 34 centimetres. During this time these tetrapods lumbered slowly across the old river deposits of southern Ireland. The animals may have been similar to to ones found in Late Devonian rocks in Greenland and probably spent a significant amount of their lives in water.

Although much of the sedimentation of the Devonian was terrestrial, in the southern part of the country marine sediments were being laid down in the Late Devonian. These rocks are locally packed with marine fossils such as brachiopods and bivalves, typical of nearshore and shallow-water environments. This marginal marine environment lay on the northern side of a new ocean called Proto- or Palaeo-Tethys, which was in the process of evolving. Ireland was now part of the old mountain belt and conditions for the formation of a new chain of mountains were already coming into existence to the south. The flooding of Ireland by a new sea, which began towards the end of the Devonian, became the dominant feature of its evolution for the next 60 million years.

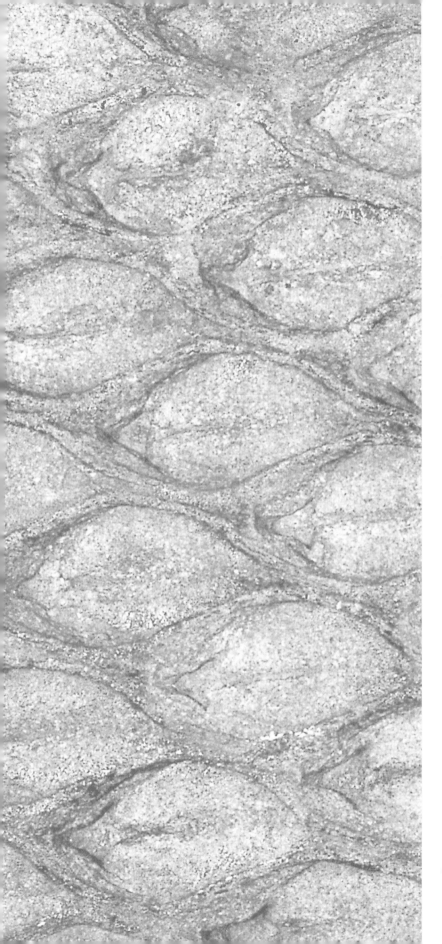

Tropical Climates

Diving in the Carboniferous Sea

Bye Bye Bahamas

The Mississippi comes to Clare

A Gentle Orogeny

The Legacy of the Carboniferous

CHAPTER FIVE

Tropical Climates

*The hills round about were covered with faded ferns like the wet hair of an ageing red-haired woman,
two grim rocks guarded the entrance to this little bay: 'Benbulbin and Knocknarea', said the driver,
as if he were introducing me to two distant relations he didn't much care about.*
Heinrich Böll. *Irish Journal*

By the beginning of the Carboniferous Period, about 350 million years ago, the relentless processes of erosion had done their work efficiently on the Caledonian mountain belt. These ancient mountains were now worn down to their roots. Much of Ireland was just above sea level and covered with the reddened detritus of Devonian sediments.

An event now occurred which, if humans had been around, would have turned Ireland into a holiday maker's paradise, an event which was to control Ireland's future shape, water resources and mineral deposits. After the traumas of the Caledonian mountain building, the piece of the earth's crust containing Ireland was allowed to relax. It began to subside in relation to sea level. The sea was therefore able to gently flood the land surface. At this time Ireland was located at about 10 degrees north of the equator, a position similar to that of the Bahamas today. So the Carboniferous seascape would have been like modern shallow subtropical seas; a fact confirmed by the study of the rocks and fossils themselves.

Diving in the Carboniferous Sea

Hutton's axiom of geology, 'The present is the key to the past', is especially applicable to our understanding of the Carboniferous. Much of the lower half of the Carboniferous stratigraphy in Ireland consists of limestones. This rock is dominated by the mineral calcium carbonate which unlike many substances, is more soluble in cool than in warm water, so it is more likely to be laid down in warmer waters. Moreover, many organisms such as corals secrete this mineral from their soft parts to produce a protective and supporting skeleton. In order to precipitate large amounts of calcium carbonate, therefore, we need an abundant and diverse animal population in a warm, and therefore shallow, marine environment.

Although the composition of these limestones is fairly simple, they often contain minerals other than calcium carbonate including dolomite (calcium-magnesium carbonate) and gypsum (calcium sulphate). These are often formed by the evaporation of sea water so they are called evaporite minerals. Their presence in the rocks confirms that the Carboniferous sea was extremely shallow in places so that the tropical heat was sufficient to cause rapid evaporation resulting in their formation. In many places within the limestones another rock called chert occurs. This rock is extremely rich in quartz (silica) and so is much harder than the surrounding limestones. These Carboniferous cherts probably formed after

Fig 5.1 Continental positions at the beginning of the Carboniferous Period, about 350 million years ago.

their host limestones were deposited; in other words they are replacement rocks. Any silica present in solution would have precipitated out at sites within the newly deposited limestones. One major source of such silica would have been the remains of sponges whose soft parts are supported by spikes or spicules made of silica. We often find, for example, discontinuous layers of chert concentrated along the contact of one limestone bed with another. In some places even the fossils have been

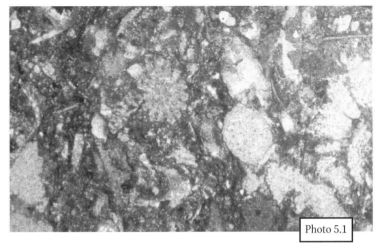

Photo 5.1

completely replaced by silica which faithfully mimics the original structure of the animal. The cherts layers or nodules may be recognized by their dense blue-grey colour and extreme hardness. This particular rock type proved very useful to the early human settlers in Ireland in the manufacture of tools and weapons such as scrapers and arrow heads.

Photo 5.1 A photomicrograph of Carboniferous limestone, possibly the most common rock type in Ireland. Most of the grains visible are the fragments of organisms including algal mats, corals and bryozoans (small colonial marine animals). Field of view is about 0.3 centimetres across.

This sea gradually spread northwards during the Early Carboniferous. The oldest Carboniferous rocks in many areas of western Ireland consist of sandstones and conglomerates deposited by rivers and on beaches near an ancient coastline. This example of a fossilized shoreline becomes younger as we move northwards in the Carboniferous. Eventually virtually the whole of Ireland became covered in a considerable thickness of limestone which is preserved today mainly in the central parts of the country, underlying vast areas. Perhaps the best known area of Carboniferous limestone is the Burren of County Clare. Here the almost horizontal beds are seen stacked apparently endlessly on top of one another, attesting to the long-lived stability of the shelf in this area. The action of water on the soluble limestone has given rise to the strange, bare landscape of this area known as karst topography.

This warm Carboniferous sea was an ideal environment for colonization

Fig 5.2

Fig 5.2 An impression of a typical sea floor scene during the early Carboniferous.

by animals and plants. A rich, abundant benthos (bottom dwelling community) existed on the well-lit seabed. A diver would have swum over communities of giant brachiopods lying on the carbonate mud, filtering nutrients and surrounded by forests of crinoids, known as sea lilies. Occasionally small mounds (bioherms) of tabulate and rugose corals would be seen punctuating the sea floor. Around them starfish and gastropods, a type of snail, would manoeuvre to feed on the shellfish and the nutrient-rich detritus. Occasionally our diver would catch a glimpse of a trilobite on or near the floor. The fauna was not restricted to the seabed however. Goniatites and nautiloids would have patrolled the water column above the rich tangle of life beneath, searching and scavenging for food. A glimpse over the shoulder by this imaginary diver would have revealed that primitive forms of shark were taking an interest in a potential lunch.

The early Carboniferous shelf was obviously stable enough to allow the slow accumulation of a considerable thickness of limestone, but there is evidence that from time to time the environment became distinctly inhospitable. Occasionally thin mudstone horizons are found interbedded with the limestones. Chemically these are very similar to the composition of ash and dust ejected from some modern volcanoes. Just south and east of Limerick city, volcanic rocks of this age are exposed which indicate significant eruptions of volcanoes sometimes standing above sea level. It seems that life on the balmy sea floor was periodically interrupted by dramatic volcanic explosions around what is now Limerick, scattering ash and dust over much of the area of Ireland. These volcanic outpourings signalled the beginning of the end of this stable shelf on which the limestones accumulated.

Bye Bye Bahamas

Sedimentation during the later part of the Carboniferous was quite different. The instability of the shelf, first marked by the volcanoes around Limerick, now became more pronounced: the shelf began to subside and the sea to deepen. This was not a regular, uniform process, however. Parts of the shelf subsided more slowly than others, giving rise to a series of 'highs' and 'basins'. The first effect of this process is seen in the deposition of the Clare Shales. The amount of deposition on highs was much smaller than that in the basins, even though the shelf as a whole was subsiding. In northern Clare for example, situated on the Galway High at the time, the Clare Shales are only about 5 metres thick, whereas farther south near the Shannon estuary, in the Shannon Basin, they are over 50 metres thick.

The presence of these very dark shales shows that the floor of the shelf was now too deep, or restricted by some barrier, for the formation of limestones. Currents were unable to affect the accumulating mud. This view is confirmed by the presence of the mineral pyrite in these rocks. Pyrite is iron sulphide. It is a bright brassy yellow mineral when fresh and because of this is sometimes known as fool's gold. Its presence means that the environment in which it formed had little available oxygen to convert the iron present into iron oxides and so it was converted to the sulphide instead. In a deep-water environment, or one which only allows restricted currents or wave activity (like the Black Sea) oxygen is scarce. Occasionally the Clare Shales contain layers rich in phosphate minerals. The formation of these minerals requires an organic rich environment. It may be that these accumulations were the result of a process known as upwelling, in which colder, deeper waters which are nutrient-rich may rise to the surface leading to the explosive growth of large masses of plankton. Upwelling may lead to a high rate of death amongst the marine population of an area because of these plankton blooms. So the combination of high mortality rates with the development of plankton and a reducing environment on the sea floor can lead to the preservation of organic phosphates.

Only a very sparse benthos could tolerate the sulphurous stinking black muds that accumulated at this time. Fish teeth and their debris contributed to the large amounts of phosphate at the base of the Clare Shales. The fish almost certainly lived somewhere in the water column rather than near the unpleasant seabed. The carbonate concretions, or bullions, typical of these shales are often packed with the shells of goniatites, which were swimming molluscs with coiled shells which swam by a form of jet propulsion, squirting fluid into the surrounding water. Like their close relatives the squids, cuttlefish and octopuses, the goniatites were probably voracious carnivores feeding on other invertebrates and fish living in this sea. Higher in the Clare Shales the sediment is less black and here there are densely crowded shell beds. The fossils are dominated by large pecten-like bivalves which probably rested periodically on the seabed between swims. They propelled themselves through the water column by the flapping of their two shells, much like scallops today.

Near the top of the shales, in a sandstone, a single specimen of one of the oldest flying insects was discovered in 1993. The insect had probably not strayed very far from land but had fallen into the sea and became preserved in a thin turbidity flow.

The Mississippi comes to Clare

This environment of quiet accumulation was now overwhelmed as large delta systems began to encroach onto the limestone-shale foundation. Deltas today are sites of massive rates of sedimentation and these Carboniferous deltas were no exception. When sediment is deposited very rapidly it may become gravitationally unstable. The layers of sediment may even become distorted and folded on, or near, the surface. These folds are known as slump folds and

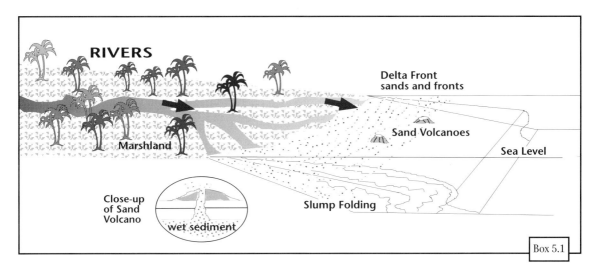

Box 5.1

Box 5.1 Deltas

Deltas are sites of enormous rates of sedimentation. The Amazon River today transports about 10 million cubic metres of sediment to its mouth every second. The arrival of such amounts of sediment on the delta front can cause instability. Frequently layers of sediments collapse and slump down the slopes, forming cascades of folds. Large amounts of water are expelled from the sediment in this process and may be forced upwards and escape explosively, forming sand volcanoes on the sediment surface, like those preserved in the Carboniferous rocks of Clare. Because of all the nutrients washed in with the sediments, deltas can support a rich fauna, but it is a hostile environment. A sudden flood in the supplying rivers can quickly bury bottom-living organisms with sediment. In the same way as humans exploit the rich soils around active volcanoes, these delta animals take a calculated risk in inhabiting this environment.

spectacular examples can be seen in exposures on the shores of County Clare around Doolin.

Another expression of this rapid sedimentation is the formation of sand volcanoes which may also be seen in this area. When sand is buried rapidly by later sediment it may still contain large amounts of water which is temporarily trapped. When the pressure reaches a critical stage this water escapes violently upward bringing the sand up with it through the overlying sediment in exactly the same way as a volcano erupts. The sand is then blown out onto the surface in the shape of a cone with a central feeder.

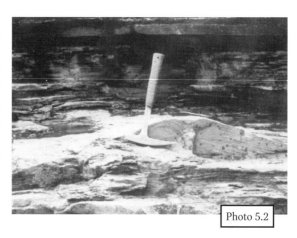

Photo 5.2 A fossilized sand volcano in the Upper Carboniferous deltaic sedimentary rocks of County Clare.

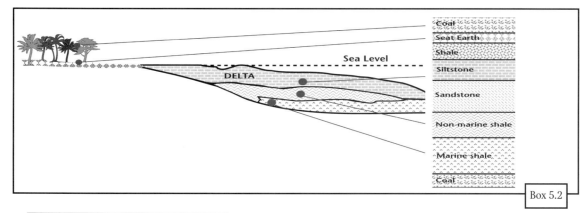

Box 5.2 Cyclothems and coal

A cyclothem is a sequence of beds which form an association. This association may be repeated many times in a stratigraphic sequence. The Upper Carboniferous in Ireland and elsewhere has well developed cyclothems resulting from deposition by deltas. Before a tongue of the delta arrives at an area the sediment consists of marine mud with a marine fauna. As sediment from the delta begins to be deposited a series of lagoons may form containing a fauna suited to non-marine or brackish water conditions. As the delta pushes seawards coarser detritus is deposited rapidly, resulting in the accumulation of sand and silt containing plant and tree fragments derived from the swamps farther inland. The channels feeding deltas frequently become choked with sediment and are abandoned. The delta then becomes inactive and its top is colonized by the marshland flora.

In the Carboniferous the marshland would have been warm and steamy, resulting in the growth of lush vegetation and huge primitive trees. The Seat Earth is a fossilized soil in which these plants grew. When the plants died they began to form peat in the wet environment of the swamp floor. This peat was buried over millions of years by further sediment until eventually it was converted to coal. Many cyclothems in the Carboniferous are incomplete and one or more elements may be missing. The diagram shows a full cyclothem, which is rare in the Irish Carboniferous. As channels elsewhere in the delta become choked, new channels may arrive at the original position and the whole process is repeated to give a number of cyclothems at any one point.

One interesting fact about the Clare deltas is that they appear to have come from the west. This would mean that at this time in the Carboniferous, a land area lay to the west of Ireland capable of supplying large amounts of sediment to the Irish basins.

Two quite distinct fossil assemblages are found in the Clare delta systems. The marine shales found at the base of a typical deltaic cycle are often packed with goniatites which lived in saline waters, seaward of the delta complex. These shells are usually flattened because the original muds have been squeezed and compacted by the overlying delatic sands. As the channels of the deltas built out into the sea however, their currents washed plant material, including the odd tree trunk, onto the delta top. In between these channels, in the nutrient-rich muds, anonymous molluscs or worms grazed, leaving behind their feeding trails, now beautifully preserved in the Liscannor Flags. These flagstones are used extensively as decorative stones in facing or paving and even for garden furniture.

Photo 5.3 The Cliffs of Moher, County Clare. These Upper Carboniferous rocks show the rhythmic cyclicity typical of sedimentation in many deltas.

A Gentle Orogeny

At the end of the Carboniferous Period Ireland experienced the effects of another mountain building episode known as the Variscan orogeny. Unlike the previous Caledonian orogeny however, there was only relatively minor deformation over much of the area. The reason for this is that Ireland occupied a part of the crust well to the north of the site of ocean closure and so escaped the metamorphism and massive igneous intrusions so typical of the central zones of orogens. However, the Carboniferous rocks did not escape unscathed. In County Cork, these rocks are quite strongly folded. As we move northwards the amount of deformation becomes progressively less. The Burren shows limestones and shales which have only suffered minor tilting and some faulting.

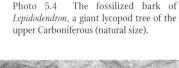

Photo 5.4 The fossilized bark of *Lepidodendron*, a giant lycopod tree of the upper Carboniferous (natural size).

One conspicuous feature of these rocks is the well-developed jointing pattern. Most rock masses contain joints, simple fractures which are often caused when the overlying pressure on the rock is relieved during uplift and erosion. As the pressure on it decreases the rock relaxes and cracks. Unlike faults, there is no relative movement of the rock on either side of a joint. Joints in limestones are especially obvious because they are fairly soluble in water, especially if the water is acidic. The joints then become enlarged and accentuated by the infiltration of water over periods of time. This

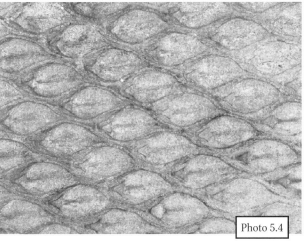

enlargement can proceed to the point where great solution hollows and eventually cave systems may develop. Limestone areas are not conducive to the formation of soils as it is often pure calcium carbonate with little or no impurities such as clay minerals to form soil. Wind- and water- borne detritus may become trapped in the joints in such areas where various plants may find refuge whilst the limestone pavement itself remains denuded of flora.

The Legacy of the Carboniferous

The Carboniferous was an important period from the point of view of the mineral wealth of Ireland. The limestones produce enormous amounts of aggregates for the building industry as well as lime for cement and for use as fertilizer on certain soils. Many major lead-zinc deposits are found in Carboniferous rocks. One example of this in Ireland is the Tyangh ore body in County Galway. The ore contained lead, zinc, copper and silver. Its formation was in part due to the collapse of the Carboniferous limestone shelf, which was accommodated by faulting as different parts of the shelf subsided at different rates. It may be that these faults acted as conduits for the injection of magma or heated fluids from deeper in the earth's crust. As these fluids reached the Carboniferous sea floor they contained solutions of lead and zinc, amongst other minerals, which were then precipitated as the solutions cooled on the seabed or within cavities in the underlying sediment.

Photo 5.5 The Carboniferous fern *Pecopteris* (natural size).

As well as influencing Irish economics, the Carboniferous Period left a heritage of landscape which influenced the settlement of the country by early man. The hard chert could be used for tools and the cave systems developed in the Carboniferous limestone afforded convenient shelter. Associated with the upland limestone scenery of the Burren, for example, is a spectacular development of karst with complex cave systems. These caves of the Burren are formed in two situations. In the eastern Burren, the percolating rain waters flow down through the soluble limestone until they reach impermeable layers of chert and volcanic mudstones. The water is then diverted laterally and forms subhorizontal caves, such as the Ailwee Caves. In the western Burren there is a line of caves just under the contact between the limestones and the overlying Namurian shales and sandstones, for example at Slieve Elva. Here the vigorous surface runoff has cascaded down the Namurian rocks and upon reaching the more soluble limestone beneath, has disappeared underground into a complicated system of caves. Such a disappearing river can be seen near the road just south of Gort in County Galway.

The Carboniferous Period was one in which Ireland saw its first profuse giant land plants atop the great deltas, a period of relative crustal stability and possibly the most idyllic climate in its entire geological history.

Photo 5.6 The gently dipping beds of Carboniferous limestone in the Burren, County Clare. These limestone pavements have little or no soil development but preserve a diverse dwarf flora in the protected joints of the rock. Many of the individual terraces are separated by thin volcanic claystones, called clay wayboards.

The Atlantic Opens

Salty Waters

An Extinction of Species

The Time of the Dinosaurs

The Chalk Sea

Did Meteorites Kill the Dinosaurs?

A New Ocean

Oil and Gas

CHAPTER SIX

The Atlantic Opens

Well, I stood, mute with shock and gazing like Cortez with wild surmise not
on some vast new ocean but at a sea of sand
that moved and tumbled, as rough as any sea I've known.
Bob Geldof

The Atlantic Ocean is one of the major features of the earth's crust; it separates the continents of Europe and Africa from those of North and South America and is clearly visible from the moon. The ocean basin occupies an area of over 100,000 square kilometres with an average depth of about 3 kilometres plunging to over 9 kilometres in the Puerto Rico Trench. The ocean system is split along a north-south axis by a sinuous submarine high, the Mid-Atlantic Ridge. Today Europe and Africa are moving away from the Americas at a rate of a few centimetres a year. Active seafloor spreading is adding new ocean crust to either side of the Mid-Atlantic Ridge. Earthquakes and submarine volcanic eruptions are common phenomena along the spreading axis. Iceland, the Azores, Ascension Island, St Helena and Tristan da Cunha, all sites of recent and continuing submarine volcanic eruptions, sit astride the Mid-Atlantic Ridge. However the development of the Atlantic basin has a long history, beginning over 250 million years ago, and intimately involving Ireland on the western seaboard of Europe.

The events in this chapter occurred during the Permian Period, the last in the Palaeozoic Era, and the Mesozoic Era which includes the Triassic, Jurassic and Cretaceous periods.

Salty Waters

Following the Variscan orogeny, which severely affected only the most southern parts of Ireland, the New Red Sandstone was deposited in a series of basins around the eroding remnants of a mountain chain that had risen during the collision of southern and central Europe. The New Red Sandstone was essentially deposited during the Permian and Triassic periods as the crust began to relax following the compression and trauma of orogenic collision. Throughout central and northern parts of Europe the Permian System commences with sandstones deposited in continental environments, called the Rotliegendes. Later, in the Upper Permian, magnesium-rich limestones, dolomites, and evaporites were deposited in the large, very saline, Zechstein sea. This distinctive system of strata was

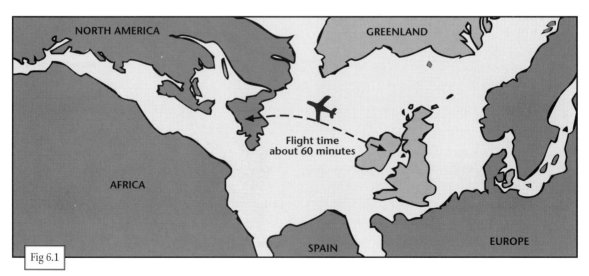

Fig 6.1. The 'Pond' about 100 million years ago before the opening of the present North Atlantic. Iceland had yet to come into existence along the Mid-Atlantic Ridge, a line of sea-floor spreading. This only shows present-day land configurations for the sake of simplicity.

first recognized in the Perm Basin of Russia over 150 years ago by Sir Roderick Murchison. There is, however, little evidence of Permian rocks exposed at the surface in Ireland, although such rocks have been found in boreholes near Larne. In the Strangford basin the Lower Permian may be represented by a thick sequence of breccias (rocks made of an angular aggregate of older rocks) with fine grained quartz sands which have the typical millet-seed shape associated with grains buffetted, abraded and sorted by wind in a desert environment.

During the Late Permian a major marine transgression swept warm, saline waters across much of central Europe. The Zechstein Sea stretched from the north of England to Poland but expansion west was largely blocked by the Pennines; however marine conditions did extend westwards through a series of straits to form the Bakevellia Sea which drowned the deserts of northwestern England and northeastern Ireland.

Conditions during the deposition of the Zechstein Magnesian Limestones were harsh. Few fossils have been recovered from these rocks in Ireland. Both at Cultra, County Down and Grange, County Tyrone the Zechstein faunas are dominated by a bivalve, *Bakevellia*, which must have adapted to high salinities. Farther east, in the north-east of England, a much richer fauna developed on the edge of the main part of the Zechstein Sea. This fauna was described by the first professor of geology at the then Queen's College, Galway, William King; his collections formed the foundations for the Natural History and Geology Museum in University College Galway, now the James Mitchell Museum.

An Extinction of Species

The Mesozoic Era has become the subject of intense research over the past few years because of its record of extinction events at its base and top. During the Late Permian many of the typical Palaeozoic animal groups were very much in decline. For example the graptolites had already become extinct during the Late Carboniferous. By the Permian only a few genera of trilobites had survived. Nevertheless the final extinction event was relatively sudden. Over 50 per cent of marine families died out in the final 5 million years of the Permian Period; during this same interval estimates

Fig 6.2 A typical sea floor during the Permian.

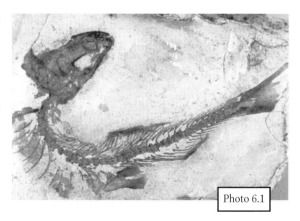

Photo 6.1 *Leptolepis*, a fossilized type of herring (natural size). These bony fish were common in the seas of the Jurassic Period when units like the Lias clays were being deposited, now preserved along the Antrim coast.

suggest that as much as 80 per cent of genera and over 95 per cent of species disappeared from the face of the earth for ever. In a sense the evolutionary clocks of many groups were reset and the earth's ecosystem was ready to regenerate with a quite different cast of organisms waiting in the wings to succeed in the new world order.

Most Palaeozoic marine communities were dominated by fixed filter feeders such as the bryozoans, brachiopods and crinoids. These animal groups dominated the shallow seas of the Carboniferous. The Mesozoic sea floor, however, was to be carpeted with mobile bivalve and gastropod molluscs as well as sea urchins and starfish more adapted to detritus feeding. The Liassic clays of Garron Point, for example, are packed with mollusc shells and the spines of echinoids.

Meanwhile on land almost 75 per cent of amphibian and reptile families disappeared. The earliest Triassic landscape was populated by clumsy dicynodonts, large herbivores filling an ecological niche similar to wild cattle. Changes in vegetation were less obvious, although there was a gradual changeover from the Late Carboniferous plant cover of various groups of fern to floras with conifers, ginkgoes and cycads .

 Why was there such a massive drop in the diversity of life-forms at the end of the Permian? Recent research suggests that a massive depletion of atmospheric oxygen occurred at that time, depriving metazoans of a very necessary substance for their metabolism.

Mesozoic rocks are relatively rare in Ireland itself, although thick sequences of Triassic, Jurassic and Cretaceous strata are located in satellite offshore basins. During the Triassic the hot and salty environments established in the Permian continued. Lower Triassic sandstones were deposited in river systems probably during flash floods generated by occasional intense rainstorms, flushing a shimmering, desert landscape. A variety of marls (calcareous muds) were deposited in large playas - temporary lakes in an otherwise hot and arid climate and some of the marlstones may be fossil desert soils. Evaporite minerals generated by high rates of evaporation are common at some horizons. Towards the end of the period a major marine transgression, an overspill from the huge Tethys Ocean which now separated much of central and southern Europe from Africa, established marine conditions again. This Rhaetic transgression brought a fauna of swimming molluscs dominated by ammonites to Ireland for the first time.

 Subsequently, during the early Jurassic or Liassic, calcareous mudstones and shales, together with some limestones, were probably deposited over much of Ireland in a shallow sea. Liassic rocks crop out in a number of places in north-eastern Ireland, for example at Garron Point, County Antrim. Most have a rich marine fauna dominated by ammonites and various other molluscs including bivalves and belemnites. Occasional bones of marine reptiles have been recovered, but these are rare. Although a possible dinosaur bone was collected from mudstones at Garron Point a number of years ago, there is little evidence to suggest that a 'Jurassic Park', like that in southern England ever existed in Ireland. Life in the Liassic seas was dominated by the carnivorous ammonites and belemnites, advanced cephalopod molluscs related to the sea squid and octopus as well as marine reptiles like the ichthyosaur and plesiosaur, and a variety of bony fish. The sea floor was dominated by deposit-feeding bivalves, gastropods and by sea urchins, in contrast to the filter-feeding benthos of the Palaeozoic. There is no evidence of younger Jurassic rocks in this part of Ireland although blocks of Middle and Upper Jurassic rocks have been identified in superficial drift deposits.

Despite the relatively poor exposure of the Jurassic in Ireland, Liassic rocks cropping out in the small fishing village of Portrush played a fundamental role in the history and evolution of geological thought. The ammonites and the hard, apparently crystalline lithology of the Portrush Rock were used by the vulcanists in the late 1700s as convincing evidence that all rocks, even those with fossils, were formed by volcanic processes. The opposing school, the neptunists, accepted an aquatic origin for all rocks. The Portrush Rock is a hard, fine grained hornfels, a mudstone which has been baked and hardened by contact with a Tertiary intrusive igneous rock. It contains a recognizable ammonite fauna and was undoubtedly once a sediment.

The Time of the Dinosaurs

The Mesozoic Era is most famous perhaps for the evolution of the dinosaurs. All the Mesozoic continents, oceans and airways were populated by large reptiles. Although the oldest reptiles are Carboniferous in age, by the Triassic various new groups had evolved. The archosaurs included the dinosaurs which were to diversify and dominate land environments during the Jurassic and Cretaceous. The saurischians or lizard-hipped dinosaurs included the giant plant-eating sauropods such as *Brachiosaurus*, *Diplodocus* and *Ultrasaurus*, weighing in at over 100 tonnes with a length of 40 metres and a towering height of 15 metres. The group also included the ferocious predatory *Allosaurus*, *Megalosaurus* and *Tyrannosaurus*. The ornithiscians or bird-hipped dinosaurs contained the armoured herbivores such as *Stegosaurus* and *Triceratops* and duck-billed hadrosaurs. The dinosaurs apparently developed a pseudo warm-blooded metabolism and certainly could maintain high rates of activity over long periods of time. The first bird, the Jurassic *Archaeopteryx* evolved from the dinosaurs; so birds such as hens and seagulls could be regarded as close living relatives of the dinosaurs.

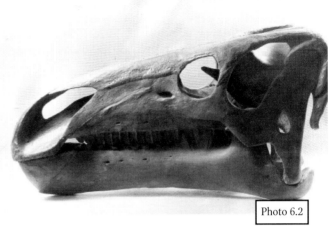

Photo 6.2 The skull of *Iguanodon*, a cow-like herbivore which roamed early Cretaceous landscapes chased by predators like *Megalosaurus*, the first named dinosaurs. The skull is about 75 centimetres long.

In the Jurassic and Cretaceous seas two main groups, the ichthyosaurs and plesiosaurs were well established. The ichthyosaurs were dolphin-like, between 1 and 5 metres in length and fed on belemnites and fish. The plesiosaurs, however, were much larger, reaching lengths of up to 12 metres. They had long necks and small heads, and by beating their large paddles they moved at relatively high speeds under water. They fed on fish, tearing them apart with their small, needle-like teeth. It is such an animal that is thought by some still to inhabit the dark, cold waters of Loch Ness in Scotland.

In the air giant flying reptiles, the pterosaurs, swooped overhead. Sadly with the exception of some reptilian footprints from the Triassic at Scrabo,

Photo 6.3 An Ichthyosaur, a swimming marine predator which plundered early Jurassic seas. Although it was a reptile, the ichthyosaur was similar to modern dolphins. This particular specimen is about 1.2 metres long.

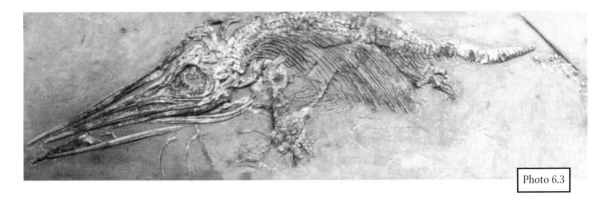

County Down, some marine reptile bones from the Lias of Collin Glen, County Antrim, and a dinosaur bone from the Lias at Garron Point, there is little evidence in Ireland of this exciting period in the history of vertebrate life on the planet.

The Chalk Sea

During this time it appears that Ireland was emerging from the Mid-Jurassic to Early Cretaceous waterways, or more likely that these rocks were sporadically deposited and have since been eroded. Whatever the reason Middle and Upper Jurassic and Lower Cretaceous rocks are largely absent throughout Ireland.

The major Late Cretaceous transgression began with a global rise in sea level. This rise swept marine conditions across much of Ireland and indeed the rest of the world. The older Upper Cretaceous rocks are sandstones, the Hibernian Greensand; these rocks are rich in glauconite and other iron minerals as well as faunas dominated by the distinctive twisted oyster *Exogyra* together with the scallop *Chlamys* and ammonites. The Hibernian Greensand was apparently deposited slowly as a condensed sequence with some disconformities. The upper parts of the Greensand occur above a significant gap in sedimentation and have different faunas, with a benthos dominated by the strongly ribbed bivalve *Inoceramus* and the sea urchin *Micraster* and brachiopods.

The most distinctive rock of the Cretaceous, the chalk, gives the period its name. The chalk occurs above the Greensand. It consists of pure white limestones made up almost entirely of calcareous disc-shaped microfossils called coccoliths and a few foraminifera. Chalk was widespread throughout Europe during the late Cretaceous. In northern Ireland the chalk seas deposited the White Limestone which has less than 2 per cent clay. This limestone is harder and more impervious than the typical Chalk of southern England, probably because of a different history of burial

Fig 6.3 A sea scene in the Jurassic

after deposition. Nevertheless like the typical chalk, nodules and sheets of flint occur at many horizons within the White Limestone.

Conditions in the chalk seas were probably unique. Coccolith ooze gathering on the sea floor provided a soft bed for burrowing animals such as sea urchins, starfish and crustaceans; fixed benthos such as the brachiopods, some bivalves, like *Inoceramus* and sponges probably needed other shells or hardgrounds, where the sediments were partly lithified and hardened, for attachment. Mobile bivalves such as the scallop *Aquipecten* often swam above the seabed between periods of relaxation on the soft ooze. Swimming above the snow-like sea floor ammonites and belemnites were largely predatory and mixed with a varied fauna of fish including sharks and skates.

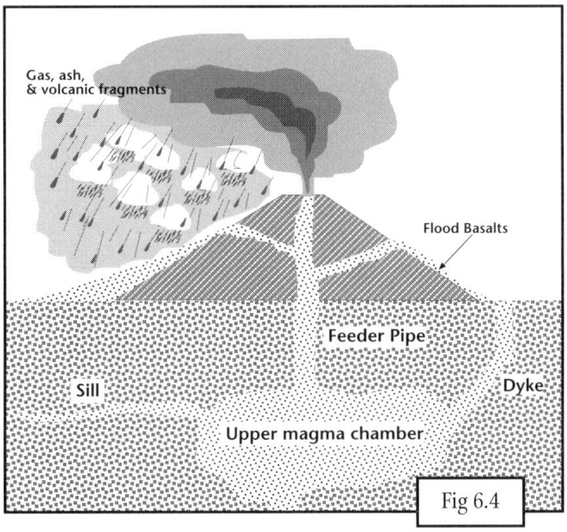

Did Meteorites Kill the Dinosaurs?

Of all the major mass extinctions, the end-Cretaceous, or K-T event, has attracted the most attention, probably because, amongst other groups, the dinosaurs met their nemesis. In 1980 Luis Alvarez and his team discovered anomalously high levels of the mineral iridium in clay at the Cretaceous-Tertiary boundary at Gubbio in northern Italy. This element is rare on earth but relatively abundant in material from meteorites. The presence of this iridium layer, together with quartz showing evidence of extreme stress, has now been described from boundary sections in many parts of the world. It may provide evidence for an extinction event driven by the impact of an extraterrestrial asteroid. A large crater, which may match the predicted end Cretaceous impact, has been discovered at Chicxulub on the Yucatan Peninsula in Central America.

However huge volumes of magma were also extruded around the time of the K-T boundary from volcanoes in north-western India, and were responsible for the Deccan Traps, enormous lava flows covering wide areas. Counter-arguments suggest that both iridium and shocked quartz may be generated during violent volcanic eruptions. Both asteroid impact and excessive volcanicity would generate thick dust clouds, block out light and simulate a nuclear winter. Total collapse of the food chain would soon follow. Many important groups disappeared for ever including the dinosaurs, sea lizards and flying reptiles, together with the ammonites and many other groups of molluscs. The scene was set for the rise of the mammals.

Fig 6.4 The nomenclature of volcanic rocks. Dykes intrude the crust vertically, sills horizontally. The ejecta from the vent of a volcano form the rocks called tuffs while lava flows form basalts. Most rocks melt between 800 and 1200 degrees centigrade.

Photo 6.4 Early Jurassic ammonites approximately natural size. These shelled molluscs, related to the living *Nautilus*, octopuses and squids, dominated the Mesozoic seas. They were voracious carnivores moving swiftly through all depths in the water column. Ammonites evolved rapidly and were widespread, making them ideal zone fossils.

Photo 6.5 The Giant's Causeway. Columnar jointing on the Grand Causeway in Antrim. The columns here are slightly tilted from the vertical.

A New Ocean

The Tertiary witnessed some dramatic changes on the western seaboard of Europe. The long history of passive crustal tension during the Mesozoic in north-western Europe was ended abruptly with the rifting of North America and Greenland from Europe. Between 65 and 55 million years ago initial updoming of the crust was followed by the separation of Greenland from Europe as sea floor spreading pushed both continents away from each other. As the North Atlantic opened, new oceanic crust was created to floor the basin as magma rose through a series of rift-parallel conduits trending roughly north-south.

The spectacular, eruptive products associated with the opening of the North Atlantic, the Antrim Basalts, are mainly restricted to north-eastern Ireland although patches of lava occur at Carlingford and Slieve Gullion. Eruptions commenced with explosive activity spreading a carpet of volcanic ash, falling from clouds of gas, tephra and water vapour, over the cracking crust. The eruption of huge volumes of basic volcanic lava followed to construct the lower part of the basalt plateau. Eruption was almost certainly from a number of volcanic centres. Near its source the lava was mobile with a ropy, so called 'pahoehoe' texture. As the lava travelled farther chemical changes and cooling took place. The lava became more viscous and developed a rough, sharp or 'aa' texture.

Red lateritic soils developed between major eruptive phases. Such terra rossa soils are common today in the tropics where rainfall is heavy and seasonal chemical weathering and leaching is intense. These processes leave only the more resistant iron minerals giving the soil its reddish colour.

The middle part of the Antrim Basalts displays the spectacular columnar jointing characteristic of the Giant's Causeway. Red lateritic horizons divide this phase of volcanism from the older and younger events. The Upper Basalts are much thicker than the lower flows, with fewer soil horizons. The flows were viscous with an aa texture and obvious flow banding. The Giant's Causeway, with thick lava flows ornamented with spectacular columnar jointing, is one of the most impressive geological localities in Europe and has featured in a number of famous paintings. Molten lava accumulated in a deep pond possibly between the volcano itself and the so-called Highland Border Ridge to the south. The slow cooling of this large volume of basic magma allowed columnar jointing as the magma contracted. Perfect columns with polygonal cross sections developed in the lower part of the pond. The Causeway itself is part of the lowest columnar zone. A number of structures have been described including regular columns comprising the colonnade with a rubbly top or entablature. These structures are extravagantly developed in the Giant's

Photo 6.6 Giant landslips on the Antrim coast showing the white Cretaceous chalk overlain by the Tertiary lavas.

Organ, where the columns appear like large organ pipes.

A series of major magma intrusions were associated with Tertiary igneous activity in Ireland. Not all the molten magma generated and poised to erupt through cracks and fractures actually makes it to the earth's surface. A great deal of magma cools and crystallizes within the crust as a variety of intrusions. A number of elongated volcanic plugs, like those at Slemish and Trosk, may have fed at least some of the Antrim volcanoes. The Killala Gabbro, exposed in Killala Bay, northern Mayo, is a giant intrusive sheet of basic and ultrabasic coarse crystalline rock. It may be associated with a series of basic dykes (vertical intrusive sheets) along the Sligo and north Mayo coast. Tertiary basic dykes have also been recorded from counties Donegal, Galway and South Mayo. They may be part of a larger dyke system associated with crustal extension. The granite plutons (very large intrusive bodies) which form the spectacular Mourne Mountains in County Down are coarse crystalline acid igneous rocks. It is possible that the granites developed by fractionation of a more basic magma. In fractionation the lighter acid minerals were separated from the main, heavy part of the magma to form granitic bubbles rising through the upper crust to form the plutons.

Photo 6.7

Despite the dramatic opening of the North Atlantic and its associated catastrophic igneous activity, some Tertiary sediments were deposited in a few basins in Ireland. The Lough Neagh Clays postdate the Antrim basalt lavas. They were deposited in a subsiding basin, on the site of the present Lough Neagh. They are rich in plant debris and freshwater molluscs. The lowest clays have thick beds of lignite derived from fossil wood and plant material. Elsewhere in Ireland Tertiary sediments are less well developed. Nevertheless uplift during this period probably initiated the evolution of Ireland's contemporary landscape. Later in the Tertiary, global cooling destroyed the warm temperate forests which hitherto were widespread.

Oil and Gas

The cracking open of the crust to form the Atlantic Ocean was to have a substantial effect on the economy of Europe. This crustal extension developed a number of subsiding basins which accumulated significant thicknesses of Mesozoic sediments and which became prime sites for the generation of petroleum. The developed world is now almost entirely dependent on petroleum products for transport energy and partly dependent on them for heating energy; until a safe and acceptable alternative is developed the ongoing search for these natural and non-renewable resources

Photo 6.7 A photomicrograph of a rock containing hydrocarbons. The tarry oil (black) is held in the tiny pore spaces between the grains of the sandstone. Field of view is about 0.3 centimetres.

is of crucial importance. But the origin of these organic compounds has been the subject of some debate. Some types of meteorites (carbonaceous chondrites) have been found to contain hydrocarbons. This might suggest that oil and gas have an origin deep in the earth and were incorporated into the earth's makeup as it formed; in other words they have an inorganic origin. On the other hand this evidence could equally be used to suggest the presence of extra-terrestrial life whereby organic carbon was responsible for these meteoritic hydrocarbons. The vast majority of oil and gas finds, however, are hosted in sedimentary rocks and it is accepted by the majority of geologists that these hydrocarbons have an organic origin. When plants and animals die their organic matter is converted to water and carbon dioxide by a process of oxidation. Obviously such a process requires oxygen to proceed. However if there is not enough oxygen the organic matter will be preserved. Such a situation will arise on the floor of a deep lake or on continental shelves which become oxygen-starved due to a high rate of organic productivity. This organic matter settles into the dark muds of these oxygen-poor environments and if these muds are buried by later sediment the organic material is converted first to a substance called kerogen and finally, with deeper burial, to oil and gas. Rocks in which the hydrocarbons formed originally are called source rocks.

As these rocks are compacted by burial, the hydrocarbons in them will be forced out of their very small pore spaces and escape, usually upwards. If there is no geological structure to trap them they will eventually seep onto the surface of the earth and be lost. In oil-fields however, the hydrocarbons have migrated into what are called reservoir rocks; rocks like sandstones and limestones which have a high porosity, that is a large number of tiny holes in them. The oil and gas seep into the microscopic pore spaces of these rocks. In order to keep the hydrocarbons in the reservoir the rock must be capped by a trap rock, that is a rock which is impermeable and will not allow the oil and gas to migrate out of the reservoir.

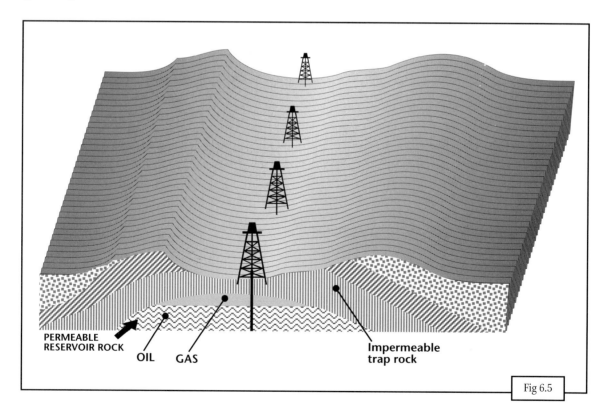

Fig 6.5 An anticlinal trap for hydrocarbons. These structures are sometimes dome-shaped, as in the highly productive oil fields of the Middle East.

In most cases it can be shown that the rocks in oil-producing fields have been buried to between 2 and 4 kilometres during their history. This would imply that the best temperature for the conversion of organic matter to oil and gas is about 120 degrees. Any intense deformation or metamorphism of these rocks would allow the hydrocarbons to escape through faulting or burn them off through heat. So most producing fields are situated in fairly young rocks (with a high organic content), gently deformed, and away from the centre of a mountain-building event with its attendant metamorphism. The rocks in the Mesozoic basins of the North Sea and off the Irish coasts fulfil these requirements. Figure 6.7 shows the sort of geology one might expect in one of the Irish offshore basins. There is undoubtedly oil or gas present in many of the Irish basins (Figure 6.6), but those off the Atlantic coast which are the largest, are also situated in the deepest waters subject to the full forces of the ocean and the cost of extraction is many factors higher than in the North Sea fields. Yet, with continuing exploration and the inevitable increase in the price of oil over the next decade or two, the Atlantic fields may become economically viable.

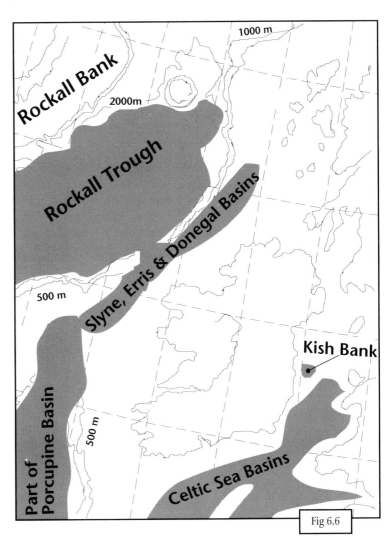

Fig 6.6 The location of basins infilled with Mesozoic rocks around the Irish coast. All of these are potential hydrocarbon targets.

Fig 6.7 The geological scenario in a typical Irish offshore basin. Depth to base is about 6 kilometres. U/C means unconformity, F means fault.

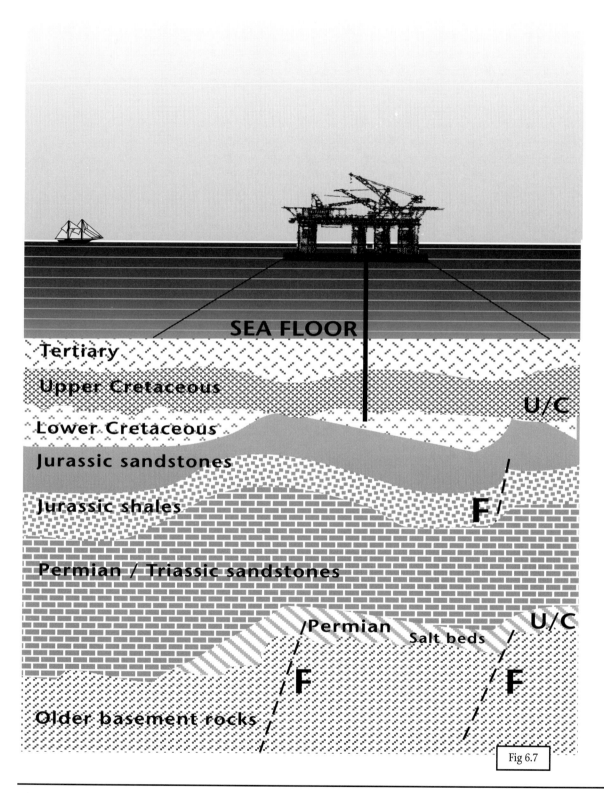

Fig 6.7

The Ice Man Cometh

The Irish Glaciers

Causes of Ice Ages

Irish Animals in the Pleistocene

The Earliest Irish

CHAPTER SEVEN

The Ice Man Cometh

History goes on;
On the rock the lichen
Records it: no mention
Of them, of us.
R.S.Thomas, Collected Poems (1995)

The Irish Glaciers

In eighteenth century Europe people believed the Great Flood was responsible for the formation of thick deposits of sand and gravel, and also for the actual rocks exposed on the surface of the earth. Primitive stone tools found in some of these gravels were held to prove the presence of an early society wiped out in the flood, the only survivors of this flood were thought to have been Noah, of biblical fame, his various relatives, and the animals saved on the ark. In 1837, however, a Swiss scientist, Louis Agassiz, who originally believed in the flood hypothesis, showed that many of the features of the landscape of northern Europe, and some of the most recent deposits could be convincingly explained by the action of thick glaciers having moved across the land surface. By studying present-day glaciers it is possible to recognize many glacially formed features preserved in the Irish landscape.

The Pleistocene Period is sometimes called the Ice Age although it was not the only major ice age to affect the earth, it was the most recent, lasting from 1.6 million years to about 10,000 years ago in Europe. The Ice Age was not a period of unrelenting cold with a permanent cover of thick ice masses. There were times of warming of the climate interspersed with the advance of the glaciers. There were warm stages and cold stages, glacial and interstadials. Some people believe that we are now living in one of these warm stages which will eventually be replaced by the re-advance of the glaciers. This, of course, assumes that we do not destroy the atmosphere before the end of this warm stage.

The debris that glacial ice contains has very erosive properties. A glacier may begin its development in an upland area. The scouring action at its base generates a scoop-shaped hollow known as a corrie. Many Irish mountains like Mount Brandon in the Dingle peninsula still preserve these corries and some are now occupied by lakes such as those in the Wicklow Mountains.

As the glacier accumulates more ice, it begins to move downslope. In this process it may exploit pre-existing river valleys. It wears the base and sides smooth to create an U-shaped valley. The erosive powers of minor glaciers feeding the main one are not so great. These minor glaciers are left occupying hanging valleys above the floor of the main valley. The enormous amounts of debris carried by glaciers in the form of boulders, sand and mud scour the underlying bedrock, forming scratches or glacial striations. By measuring the directions of these and other features we can tell the way in which the vanished glaciers moved. Roches moutonnees are such directional glacial structures. They are exposures of bedrock which have a long smoothed surface dipping in the opposite direction to a shorter rough surface. The rough, hackled surface dips in the direction of ice movement.

When glaciers melt, all the detritus they contain is dumped in an unsorted mixture with a wide range of grain size. This material is often in the form of boulder clay, which may be deposited from the base of a glacier as it moves, or in the form of a terminal moraine, a long transverse mound of sediment dumped at the maximum point of advance of the glacier. Eskers, which are very common in the midlands of Ireland, are sinuous mounds of sand and gravel which may represent the deposits of meltwater rivers which flowed within the glaciers themselves. When the glacier melts these subglacial river deposits are left as linear upstanding mounds which may originally have extended for many tens of kilometres. With the retreat of glaciers large blocks of ice may be left stranded. The melting of these blocks leads to almost circular depressions in the ground surface which remain filled with water. These are the kettle-hole lakes which are preserved in many low-lying areas of the country.

The early glacial record is absent or poorly represented in Ireland. In all likelihood cold conditions became more common in Europe some 2 million years ago, but in Ireland the earliest record of such conditions is much later, in the Pre-Gortian (Figure 7.1). Sediments of this age are found, for example, in Gort, County Galway, as the deposits of a glacial lake and contain the pollen of herbs and dwarf plants, which is taken to indicate the ending of a cold stage . The subsequent Gortian sediments, peat and mud, contain a much greater diversity of plant material indicating that at this time Ireland was colonized by juniper, alder, ash, hazel, oak and yew.

The rhododendron was also present. This is not the natural ancestor of the plant we see today in Ireland. The original Irish rhododendron was pushed back to central Europe during subsequent glacial advances and the present-day plant was reintroduced to the country by people in the past two centuries.

Near the top of the Gortian sediments the pollen remains become dominated by those of grasses and herbs. This indicates the dying of the woodlands as colder conditions became re-established. The plant material found in Pleistocene sediments is useful not only for dating of the sediments but also as a sensitive indicator of any temperature changes in the environment.

The retreat of these ancient Irish woodlands heralded the advance of the Munsterian ice tongues. Three ice masses appear to have affected Ireland during

Photo 7.1

Photo 7.1 Glacial deposits called till, or boulder clay, within a drumlin. The lower muds near the two figures were deposited on the floor of the Pleistocene Galway Bay and contain ice-rafted pebbles. The upper part of the section was deposited partly by great surges of meltwater into the waters of the bay and partly by ice.

Fig 7.1 The timescale of the Pleistocene glaciations.

Fig 7.2 The maximum distribution of Pleistocene ice in Europe and Ireland.

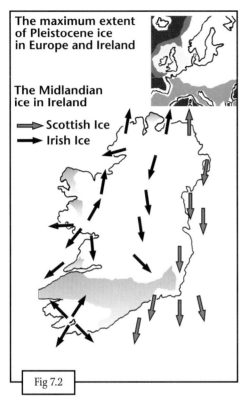

this cold stage. The first glaciers formed were in the Irish midlands. Next large masses of ice moving down from Scotland soon began to affect the country (Figure 7.2). These ice masses were funnelled down through what is now the Irish Sea basin. They affected northern and eastern Ireland as well as Wales and south-western England. A third centre of ice developed in Connemara, carrying erratic boulders of the Galway Granite as far south as Kerry.

How thick were these ice masses? Today, in the centre of Greenland, the ice is over 3 kilometres thick, and although it is unlikely that any ice masses reached these majestic dimensions in Ireland, the Connemara ice was probably over 300 metres thick even as far away as the Wicklow mountains.

The warm stages of the Ice Age are harder to recognize than the cold ones since renewed glacial advances frequently erode the sedimentary evidence of their existence. But clues do remain. For example there is about 90 centimetres of white sandy mud found at Baggotstown, County Limerick, lying between two horizons of boulder clay which suggests evidence for the Ipswichian, or Last, warm stage in the form of lake deposits.

A return to cold conditions began about 70,000 years ago in Ireland and the rest of north-western Europe. This stage is called the Midlandian in Ireland since the major ice developments this time were centred in the midlands of the country. There were a number of advances and retreats of the ice during this stage. Many of the geographical features of the midlands are due to the passage and melting of these glaciers. They developed the esker and drumlin morphology so typical of this part of the island. Some highland areas such as Benbulbin in County Sligo, rose higher than the level of the ice. These upland areas, known as nunataks, managed to preserve an arctic-type flora through this time. The ice was in retreat by about 15,000 years ago and

Photo 7.2 Benbulbin, County Sligo. A great ocean liner of Carboniferous limestone sailing the sea of the bogs. This mountain was probably a nunatak during the latter phases of the glaciation, that is a mountain whose top rose above the level of the ice and so escaped the worst effects of the scouring.

woodland became progressively better established.

Major glacial events have a dramatic effect on sea level. In the formation of glacial centres much of the water of the earth is locked up in the ice. Thus sea level will tend to fall. However, the weight of the ice can also cause depression of the crust. So there is a constantly shifting balance between subsidence of the land surface and changes in sea level. When the ice sheets melt, water is released back to the oceans and sea level rises. This is followed shortly by uplift of the land in response to the removal of the weight of the ice. During a major glaciation then, sea level changes can be complicated. There is much evidence in Ireland to demonstrate these changes.

In Clew Bay, for example, there are numerous drumlins which were formed on land during the Midlandian cold stage. These are now partially drowned by the sea showing a sea level rise since about 30,000 years ago. On the south side of Clew Bay is a man-made stone circle. These circles are typically about 5000-3500 years old. This circle is now situated in marshland which is occasionally flooded by the sea, evidence of further sea level rise since 5000 years ago. Numerous drowned bogs around the Irish coast also attest to sea level rise since their formation. Conversely the presence of raised beaches at many places show how much the land has risen since the last cold stage. At Malin Head, for example, the raised beach is some 20 metres above present sea level. So, the relationship of land to sea is complicated during glaciation. Yet we can make some estimates of past sea levels with respect to that of the present day (Figure 7.3). The lowest sea level during the Pleistocene was around 130 metres below what it is today.

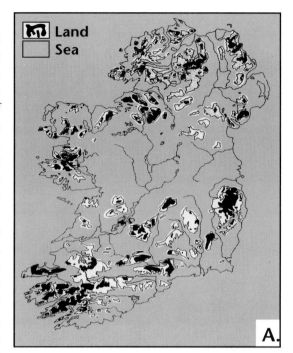

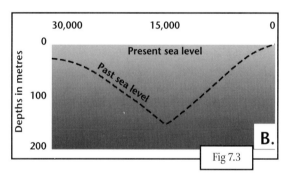

Causes of Ice Ages

The causes of major ice ages are not yet fully understood. There is no universally accepted theory but there are some points which may be important.

The sun's energy varies from time to time. In the short term these variations are visible as sun spots and flares off the solar surface. It has not been proven however that such regular variation in the amount of energy the sun gives out could explain a global cooling episode.

Major volcanic eruptions release huge amounts of gas and dust into the atmosphere. This prevents much of the sun's radiant energy from reaching the surface of the earth. There is as yet no evidence however that any of the main glaciations to affect the earth have been preceded by unusual amounts of volcanic activity.

The position of continents may influence the amount of heat absorbed by the earth. Continental areas reflect a lot of the sun's energy whereas oceans absorb heat and act as heat sinks for the planet. If large continental areas become

Fig 7.3 (a) What Ireland would look like if the sea level were to rise by about 150 metres; (b) The changes in sea level with respect to the present day over the past 30,000 years.

concentrated around the poles through the processes of plate tectonics, these large cool areas may affect ocean currents and heat distribution. This is especially true if continental masses are separated by narrow, deep oceans which do not have large surface areas exposed to the radiant energy of the sun. These configurations of continental masses may result in a gradual cooling of the earth's surface.

Irish Animals in the Pleistocene

Very few animals in Ireland would have survived the glacial advances, but in the interstadials there is evidence of significant colonization not only by the plants already mentioned but also by insects and vertebrates. There were two main phases of vertebrate occupation in the Pleistocene, knowledge of which has been obtained largely from the excavation of cave systems. The absolute ages of these fauna have been determined by dating using the carbon-14 method which is reasonably accurate up to about 50,000 years ago.

Before the Midlandian advances, Ireland was home to the famous woolly mammoth, frozen examples of which have been recovered from Siberia. Bones of this animal in Ireland have been dated at around 40,000 years old. This type of elephant was accompanied by bear, wild horse, wolf, spotted hyena, arctic fox, reindeer and various types of lemming. Also roaming the land was the giant Irish deer carrying huge antlers up to 3 metres across. During the re-forming of the Midlandian ice caps this fauna retreated to warmer climates but some returned towards the end of the last glaciation about 13000 years ago including the brown bear, the giant deer, the red deer, the wolf and the lemming. They would have crossed into Ireland across various land bridges during sea level falls (Figure 7.4).

As the climate began to deteriorate some of them, such as the giant deer became extinct. The Pleistocene fauna in Ireland is impoverished in comparison with Britain and Europe owing to the fact that much of the country was completely covered in ice from time to time. Fossils of this age from southern England, which was probably never severely glaciated, include the bones of lion, hippopotamus and rhinoceros. Fortunately this feature also kept the snake out of Ireland!

The Earliest Irish

Humans lived in England and Wales prior to the great glacial advances of the Pleistocene. The Swanscombe skull from England may be around 320,000 years old and a tooth from north Wales about 200,000 years old. Human settlement in Ireland seems to have begun relatively late. Early Europeans were able to escape southwards during glacial advances. There is some evidence, for example, to suggest that Neanderthal humans moved south to the Middle East during times of glacial advance in northern Europe.

In England there are human artefacts from Ice Age deposits which indicate the presence of humans over 25,000 years ago and actual human remains in Wales are dated at about 16,000 years ago. These people had crossed the marshland which occupied the English Channel during the Midlandian cold stage but as they have left no vestiges of long-term settlements they were probably seasonal hunters.

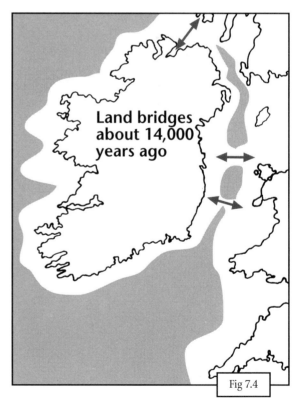

Fig 7.4 Possible land connections between Ireland, Wales and Scotland about 14,000 years ago. If these in fact existed, they were probably tidal marsh-lands rather than dry ground.

It is likely they were following the migrating herds of deer and would have done their hunting in full view of the southern margins of the great Scottish ice sheets. These hunters, part of the Palaeolithic culture, were well versed in the arts of geology. Their temporary settlement sites are preferentially developed around areas where chert and flint were available; an added bonus would have been rock formations which easily formed cave systems for shelter.

Fig 7.5 An impression of a scene during the Pleistocene. During interglacial stages humans would have migrated northwards as the ice withdrew. Fossil remains of these animals have been found in Pleistocene deposits from Ireland or Britain. From front to back these are: brown bear, spotted hyena, horse, giant deer, wolf, bison, lion and woolly mammoth.

By about 10,000 years ago the European climate had improved considerably. The last of the ice had retreated to northern Scandinavia. A new culture, the Mesolithic, had moved into Europe. The tools and hunting weapons of these peoples were better crafted than their Palaeolithic ancestors, although the two cultures would have existed side by side for a long time. There is poor evidence for a Palaeolithic incursion into Ireland but, about 9000 years ago it seems that humans arrived in the country in sufficient numbers to leave traces of their occupation behind.

What sort of country would these people have encountered? First there would have been numerous lakes and rivers in the low-lying midlands as a result of the glacial melting. Secondly, pollen recovered from sediments of this age show that the landscape would have been relatively inhospitable, comprising huge areas of forest with a tangled undergrowth of hazel beneath pine, elm, oak and alder trees. This was difficult country to penetrate and it was originally thought that these people lived as hunter-gatherers around the coastal fringes of the island. However, excavations at Lough Boora in County Offaly and Mount Sandel near Coleraine show well developed settlement sites established at inland areas demonstrating that they were not merely coastal scavengers. Indeed, there is evidence of a major cottage industry in flint tools established near Newferry in County Antrim. A layer of charcoal near Carrowmore in Sligo has given a radiocarbon date of approximately 9500 years. If this represents the remains of a settlement site it is the oldest in Ireland so far discovered.

This culture flourished in Ireland until about 5500 years ago. There then occurred a major revolution in the life style of these early settlers which was to affect the Irish landscape right up to the present day. The cultivation of land for crop growing was introduced around this time. The amount of pollen in sediments of this age declines rapidly which is thought to reflect the deliberate clearing of trees by these early farmers. It may also be linked to the spread of tree diseases coincident with the farming clearance. The killing of the trees was probably achieved by a process known as ring barking (Figure 7.6) whereby the bark was removed in a ring around the tree which eventually resulted in its death without the intensive labour investment of tree felling with stone axes. Once the shade provided by the trees had disappeared, grasses would begin to grow on which the cattle and sheep could graze. Once the cleared area had been grazed to depletion, the whole process would be repeated elsewhere, these people were not environmentally friendly!

This change was probably not an 'invasion' of Ireland. Both cultures would have co-existed for a lengthy period. The farmers would have arrived in small boats like the modern curragh, each of which would have contained seeds for the spring planting and the small amount of livestock each family

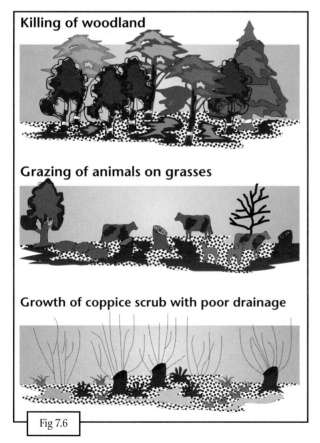

Fig 7.6

Fig 7.6 The destruction of the primeval Irish forests began with an increase in climate wetness and the growth of the bogs. The rate of loss was probably enhanced by the clearances of the Neolithic farmers over 5000 years ago.

possessed. One effect of the agricultural way of life was the establishment of permanent settlement sites. Although the remains of wooden hut structures are naturally difficult to find, these people left a more permanent legacy in the form of various stone structures, elaborate burial and ceremonial edifices such as court-graves, passage-graves and dolmens which attest to a society which was organized and with time to spare. Not for them the constant gathering and hunting of their predecessors. Rather they were able to devote considerable resources to their ceremonial structures. It is interesting to note that the great passage tomb of Newgrange was begun 1000 years before the first dated pyramid in Egypt. Some of these structures, such as the dolmens (Figure 7.7) were mainly for burial, but others were more enigmatic in their purpose. The stone circles are thought to have been a form of calendar. The alignments of the stones may have been used to ascertain quite precisely the time of year, a function of supreme importance to a people dependent on farming for their existence.

The European and Irish climate was changing even before this Neolithic culture arrived in the country. It became dominated by weather systems from the Atlantic bringing wetter conditions. At the same time the great post-glacial lakes of Ireland were invaded at their margins by water-loving plants such as water lilies and eventually reeds and sedges which reduced their size. The plant root systems trapped mud and silt which led to the filling of the lakes by sediment. The debris from the dead plants formed fen-peat. If this peat built up to a sufficient thickness the plants began to find it harder to obtain nutrients from the underlying soils and gradually fell back to be replaced by sphagnum moss, a plant that is able to survive on nutrients contained in rain water alone. The lakes were now turning into great marshes covered with sphagnum which began to impede the natural drainage of the countryside. These proto-bogs stored vast quantities of water and began to raise the water table around their margins. Drainage worsened progressively.

A further deterioration was caused by the activities of the Neolithic

Fig 7.7

Fig 7.7 The original burial/ceremonial mounds of the Neolithic culture are now only preserved as capstones supported on stone uprights.

farmers. The death of the large trees prevented their root systems from draining the soils and this allowed the spread of bog conditions into areas which were once forested. The vestiges of these great forests can be seen as tree stumps (bog-wood) revealed beneath the bogs when the turf is cut today. The removal of trees has gone on in Ireland ever since. There are few if any living remnants of the original Irish forests left on the island.

Whilst the stone age people of Ireland knew precisely where and how to extract chert from the Carboniferous Limestone, later people of the Bronze and Iron Ages were even more expert geologists. They brought with them from Europe the knowledge of the rock types and geological settings likely to contain copper, gold, tin and iron. They had developed the art of smelting the raw ores of these metals for the manufacture of tools as well as some of the most beautiful bronze and gold ornaments in the world. Here we reach the historical past where the features of the country are partly controlled by humans rather than purely by geological processes.

The geological evolution of Ireland is, in microcosm, the evolution of the earth. It shows how the collision of continents and the birth of oceans generates planetary crust. It shows how different environments have not only controlled the types of rocks formed but also the nature of the life forms which inhabited certain niches. The rocks of Ireland contain the evidence, written in stone, of all the events that have shaped the island. This rock framework of the planet on which we live controls our lives. From these rocks we obtain the hydocarbons which drive our planes and cars and provide our plastics. They control our soil types and our vegetation. From them we derive our aggregates and cement for building, our steel and copper for industry and our gold and diamonds for decoration.

The environment in which we live today is the result of processes which have gone on for almost incomprehensible periods of time. Typically oceans may be created and destroyed in about 100 million years. Major deltas, such as those of the Late Carboniferous may have lasted some 20 million years. Extensive glaciations may persist for 5 million years. The human species has so far lasted some 3 million years, a speck on the earth's clock face. But we are now the trustees of a planet which has managed perfectly well without us for over 4000 million years. Let us be careful with it.

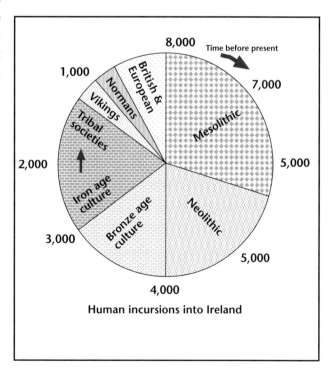

Fig 7.8 Distinctions between various cultures would have been much less distinct than the straight lines of the diagram imply, with cultural mixing rather than separation being the norm.

Fig 7.8

The Trail Guides

The South-West Donegal Trail

The Connemara Trail

The Clare Trail

The Dingle Trail

The Wexford Trail

The Antrim Coast Trail

The Trail Guides

In keeping with the aims of the book, the trail guides cover simple aspects of the geology of particular areas. They are designed so that a single trail can be completed by car in between a half and two days. None of the trails involves lengthy or difficult walking and most stops are easily accessible from the road.

It is the users responsibility to ensure that access to various stops is open to the public and that, on coastal stops, the state of the tide is safe.

Features within rocks such as minerals, sedimentary structures or fossils, are best seen close to the rock. A hand lens is useful for seeing smaller objects in rocks. Always look for 'fresh' rock, that is rock which is not overgrown with lichen or algae. Each trail requires an adequate road map such as the Ordnance Survey 1:50,000 publications. The numbers of these maps are stated near the beginning of each trail guide.

The Irish Geological Association publishes short field guides to a number of places in Ireland at a nominal cost. It also sells collections of rock samples and explanatory booklets. Anyone is welcome to join the association which runs field trips to various locations from time to time. At present, enquiries may be made to the Department of Geology, Belfield, Dublin 4, telephone 01-706 2138, or at any University Geology Department on the island.

No one trail can be said to offer a comprehensive view of the geology of an area, however the following is a list of other guides to some areas in the country. Some of these require some previous knowledge of geology.

Anderson, T.B., Hutton, D.H.W., Phillips, W.E.A. & Roberts, J.C. 1978. Field Guide to a traverse in the northwest Irish Caledonides. Geological Survey of Ireland Guide Series 3, 50pp.

Brück, P.M., Gardiner, P.R.R., Max, M.D. & Stillman, C.J. 1978. Field Guide to the Caledonian and pre-Caledonian rocks of southeast Ireland. Geological Survey of Ireland Guide Series 2, 87pp.

Charlesworth, J.K. & Preston, J. 1958. Geology around the university towns: northeast Ireland - the Belfast area. Geologists' Association Guides 18, 30pp.

Emeleus, C.H. & Preston, J. 1969. The Tertiary volcanic rocks of Ireland. IAVCEI Symposium, Oxford and Queen's University, Belfast, 70pp.

Horne, R.R. 1979. Geological Guide to the Dingle Peninsula. Geological Survey of Ireland Guide Series 1, 53pp.

Todd, S.P., Boyd, J.D., Sloan, R.J. & Williams, B.P.J. 1990. Sedimentology and tectonic setting of the Siluro-Devonian rocks of the Dingle Peninsula, SW Ireland. British Sedimentological Research Group Field Guide, 17, 157pp.

Trail Guide Number 1 - The South-West Donegal Trail

Donegal is dominated by rocks of the Dalradian Supergroup and large granite bodies forming upland scenery reminiscent of Scotland. It also contains evidence of the opening of the present Atlantic Ocean and a fossilized Carboniferous coastline.

Ordnance Survey 1:50,000 Map Nos. 10, 11 or suitable alternative.

Route: The trail can begin in the county town of Donegal. Leave on the road west for Killybegs (N56). Leave Killybegs on the road west (T72A). After 7 kilometres turn left onto a small road signposted 'Coast Road'. After about 4 kilometres turn left for Muckross Head; bear right until the road ends at a small car park on the foreshore.

STOP 1 - A Carboniferous shoreline

This stop is best at low to middle water and not a good idea during storms! About 75 metres ESE of the car park in a small cliff is an anticlinal fold in these Carboniferous rocks. An anticline is a fold whose limbs dip away from the axis; in a syncline the limbs dip towards the axis. In this case the limbs dip at a shallow angle of about 20 degrees. This fold shows the very gentle effects of the Variscan orogeny on this part of Ireland. Deformation in Carboniferous rocks increases in intensity as we move from north to south in Ireland. Walking southwards towards the ocean the rock platform shows trough-shaped cross-lamination in calcareous sandstones and quartz-pebble conglomerates. This structure indicates currents flowing from the east during deposition of these sediments, about 360 million years ago. Continuing southwards, as the cliffs begin to rise, large solitary fossilized burrows are exposed, especially noticeable on the bases of some of the beds. A dark shale bed near the top of the cliff here shows a cross-section through these burrows. Round the corner of the cliff here and looking south east at the facing cliff, channels cut by deltaic currents and now infilled by sandstones are exposed.

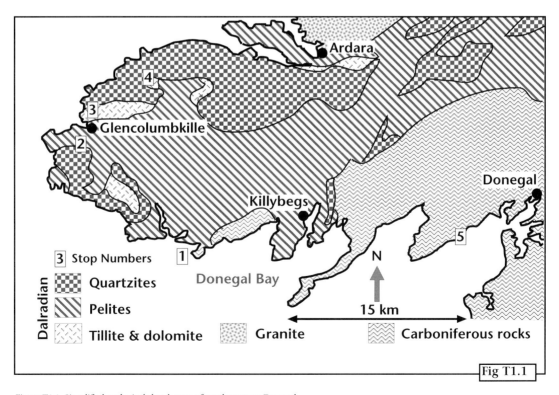

Figure T1-1. Simplified geological sketch map of south-western Donegal.

These rocks were formed early in the Carboniferous on deltas near an ancient coastline. These are not the same deltas as those mentioned on the Clare Trail (Trail 3). Limestones do not tend to form on sea floors where a lot of detritus is being washed in from the land. As the Carboniferous Period progressed, this old shore was overwhelmed by the northward-spreading sea which became established here, eventually forming the limestones so typical of the Lower Carboniferous elsewhere.

Route: Return to the coast road and turn left at the junction for Kilcar. Exposures here on the right hand side of the road are of the Termon Pelites (Dalradian fine-grained metamorphic rocks) covered by Pleistocene deposits. Drive through Kilcar to Carrick. About 4 kms NW of Carrick turn left onto the road signposted 'Mailainn Mhoir'. Five kilometres farther on stop at small lay-by with a broken black and white signpost.

STOP 2 - Ancient rites
Cross the small footbridge here and follow a narrow concrete path for about 60 m. Here is a well-preserved example of a court tomb or cairn. These are considered to be among the earliest ceremonial/burial structures in Ireland and are commonly in the age range of 5500 - 4500 years before the present. They are often elaborate monuments involving an enclosed court leading to two or more smaller chambers. The court was probably used for ceremonies associated with burial. Whilst similar tombs have yielded evidence of large numbers of burials, others often contain only fragmented bone and pottery material. This suggests that the small burial chambers were periodically cleared out to make room for new arrivals on their journey to the other world. Originally the whole structure would have been enclosed within a stone cairn or mound. This particular example has been partially reconstructed.

The tomb was constructed originally by some of the earliest agriculturalists to settle in Ireland. The final version of the tomb is so elaborate however that a considerable communal effort must have gone into its building. Thus the farming community by this time must have been well established so that members could be spared from their normal tasks to undertake the construction.

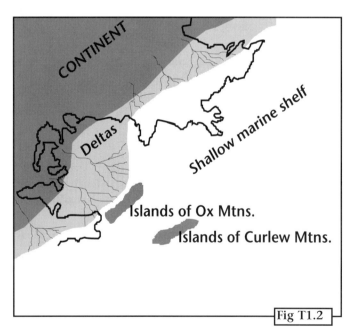

Route: Continue on the road to Glencolumkille. On the approach to the village is a fine view of sea cliffs with a tower on the resistant Slieve Tooey Quartzites of the Dalradian. This is the site of the wrecking of a ship of the Spanish Armada in 1588. Glencolumkille is an area rich in archaeological remains. Enquiries may be made locally or refer to Archaeological Survey of Donegal by Brian Lacy and others published in 1983 by Donegal County Council. Turn left at the village for Ardara. After 300 metres turn left at the road junction where there is an inscribed standing stone to your left. After 200 metres turn right onto a narrow tarmac road. After about 1.75 km pull off at a long right hand bend.

Figure T1-2. The palaeogeography of the area in early Carboniferous times.

STOP 3 - Plastic rocks

From a small quarry in glacial deposits here walk about 100 metres NNW towards the sea cliffs and follow a gentle gully to the rocky shore. These exposures demonstrate the intense folding which the Dalradian rocks have undergone. The rocks here are pelites (the metamorphic equivalent of shales) interbedded with red-brown stained quartzites. The quartzites are very hard and resistant to weathering. Large thicknesses of such rocks frequently form upstanding topography such as Errigal Mountain in northern Donegal. At the northern end of the exposure, before the high cliffs, are excellent examples of folds in these rocks. When buried in the crust at high pressure and temperature, rocks can behave plastically when subjected to stresses. Dalradian rocks have been subjected to several episodes of such folding. While these are small-scale folds they reflect the form of much larger folds in the Dalradian. Some folds may have wavelengths of many kilometres. We can envisage these large folds underlying much of Donegal by looking at these small scale equivalents. These represent much more intense pressures than the folding in the Carboniferous rocks at Stop 1.

Route: Return to the standing stone at Glencolumkille and go straight on the road signposted 'Ardara'. After about 4 km turn left and make for Lough Kiltyfanned along poor roads. The lake is situated on the narrow road to the village of Port on the coast.

STOP 4 - The world greenhouse

Stop on the road about halfway down the length of the lake, where distinctive yellow-brown rocks are exposed in a small bluff on the left of the road. These rocks are part of the Dalradian succession and about 700 million years old. They are dolomites which have been metamorphosed in common with all Dalradian rocks. A dolomite is a form of limestone but contains large amounts of magnesium. Such rocks can form in several ways but they are commonly generated in extremely shallow marine conditions in an arid climate. Evaporation of shallow washes and pools can concentrate magnesium salts from the sea water. This concentration is often enhanced by the presence of certain algae. It may be that the conspicuous banding present in these rocks represents the remains of alternating bands of algal matting and sediment. In any event these dolomites are thought to have formed in warm, clear seas in the intertidal zone at the margins of an ancient landmass.

About 70 metres SE from this bluff, near the lake, are exposures of grey rocks. These are metamorphosed sandstones. If you look carefully at clean exposures you will see that these sandstones contain pebbles and cobbles of other rocks such as dolomite and granite. This rock is a tillite, a fossilized sediment deposited from glaciers. From Connemara through Donegal to Scotland, these Dalradian tillites may be found at the same stratigraphic horizon. This is taken to indicate a significant glacial event in the Precambrian. Since the tillite directly overlies the dolomites it means that even subtropical areas suffered a rapid and drastic drop in temperature. A possible cause suggested for this has been the rapid depletion of the atmosphere by the locking up of large amounts of carbon dioxide to form the limestones and dolomites. This would have generated an 'anti-greenhouse effect'. Heat from the sun would have been reflected straight back into space without the normal reflection back to earth afforded by a normal atmosphere, thus initiating a drop in temperature.

The hills around the lake consist of the Slieve Tooey Quartzite, which is also Dalradian and overlies the tillites.

Route: Turn around and leave eastwards taking the narrow road for Ardara across typical wild Dalradian scenery. Before Ardara turn right onto the T72/N56 for Killybegs. At the major road junction turn left for Donegal town. At Mountcharles turn right around the middle of the village onto a small road signed 'Methodist Church'. Make for Blind Rock (on the road map) by turning left after 2 kilometres and left again after 1 kilometre. Follow this narrow road until it ends at a small parking area near the beach. It is best at low to mid-water.

STOP 5 - The beginning of the Atlantic Ocean

Walk west from the parking area for about 120 metres. The rocks in the small cliffs are silty sandstones of the Carboniferous, which stratigraphically overlie those seen at Muckross Head. These are deeper-water shelf sediments. They are not particularly fossiliferous but if you look carefully you may see fragments or partial stems of crinoids, especially near the parking area. So by this stage of the Carboniferous Period, the very shallow deltaic detritus of the old coastline seen at Muckross Head had been swamped by the major transgression depositing deeper-water sediment on top.

On the rocky sea platform here you will see two linear rock structures, each about 1 metre wide, striking SSE into the sea, looking a bit like breakwaters. These are vertical dykes, rocks of solidified magma which was injected through the Carboniferous rocks during a period of crustal extension. One of them can be seen cutting through the rocks of the cliff. At the margins of these dykes the country rocks are slightly altered, having been baked by the intense heat of these intrusions. The rock comprising these dykes is known as dolerite, a basic igneous rock usually with little or no quartz.

Walking about 100 metres farther westward across the beach you will find at Blind Rock a much larger dyke some 20 metres across, with horizontal cooling joints. These dykes are of Tertiary age (about 50 million years old), and are related to the opening of the present day Atlantic Ocean. As the great landmass of North America and Europe split apart to form the new ocean, hot magma from chambers within the crust was allowed to surge upwards along the cracks formed by the extension. It is possible that some of these dykes acted as feeders to flows of lava on top of the Tertiary landscape, long since disappeared in this part of Donegal but partially preserved in Antrim (see Trail 6).

Route: Return to main road at Mountcharles and turn right for Donegal town.

Trail Guide Number 2 - The Connemara Trail

The area contains a wide variety of geology and is often used by universities and colleges in geological field-work training. This trail provides a window on the evolution of Ireland in the Precambrian, Ordovician and Silurian, from about 700 to 400 million years ago.

Ordnance Survey 1:50,000 Map Nos. 37, 38, 44, 45 or suitable alternative.

Route: The trail begins in Cong village, south-east of Lough Mask.

STOP 1 - The waters rise

At the north end of Cong village is the Rising of the Waters. The Cong area is underlain by Carboniferous limestone. Owing to its solubility in water, this rock frequently forms caves and underground channels. Such areas often show sinks and risers. A sink is where a river or stream disappears underground into one of the limestone caverns. These underground rivers may rise again to the surface elsewhere, as they do here in Cong.

The unpredictability of water movement in limestones is ironically further demonstrated by the infamous Dry Canal at the eastern end of the village. This canal, constructed in the late nineteenth century, was intended to join the two great lakes of Corrib and Mask. Because of inadequate sealing of the canal floor, the water just disappeared into the cracks and caverns within the limestone beneath, leaving behind this monument to mismanagement and ignorance of the importance of geology in civil engineering projects.

Route: Leave Cong northwards on the road to Clonbur (L101). After about 200 metres, branch left. Take the first small road to the right signposted 'Pigeons' Hole'. After about 150 metres pull off to the right in a lay-by at the woods.

STOP 2 - The Pigeons' Hole

Follow the track through the woods for about 100 metres and arrive at the Pigeons' Hole. This is an opportunity to see what happens deep within the Carboniferous limestone in this area. Descend the steps carefully. The hole is part

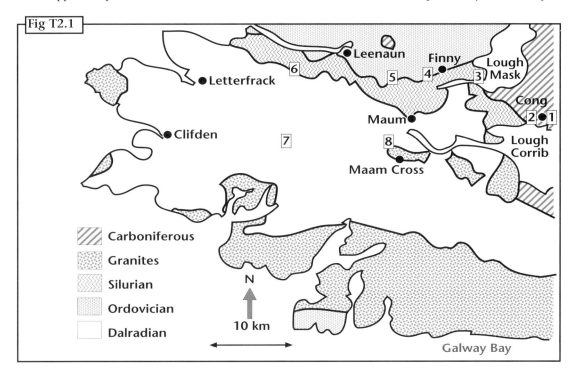

Figure T2-1. Simplified geological sketch map of the Connemara area.

of a sinkhole and cave system whose roof has collapsed (see Figure T2.2). At the bottom of the hole an underground river flows through the cavernous limestone.

Route: Continue along the road to the intersection with the Clonbur road after about 120 metres. Turn left for Clonbur here. At Clonbur village branch right for Finny. After about 4 kilometres pull off at the Ferry Bridge where there is a parking area.

STOP 3 - Uplift of the mountains

A chance here to stop and draw breath after the ascent from the Pigeons' Hole! The view from the bridge shows, to the east, the rolling lowlands of the Carboniferous limestone fringing the eastern margins of Lough Mask, and stretching more or less continuously as far as Dublin. To the north are the mountainous terrains of the Ordovician and Silurian rocks of southern Mayo. The very top of the high plateau of Maumtrasna exposes patches of rock which represent the base of the Carboniferous, unconformable on Ordovician sandstones. The fact that the Carboniferous is also exposed to the east of you here, at lake level, means there must have been considerable fault movement between these two areas in the past. These faults therefore accommodated the uplift of the Maumtrasna area by at least 700 metres after the deposition of the Carboniferous.

Route: Continue on the main road towards Finny village. The high ground to your right is made up of the hard volcanic rocks of the Ordovician and the lower slopes consist of softer Silurian rock unconformably overlying these. Continue through Finny village. About 400 metres west of the village stop at a small stand of pine trees around a cottage on the right-hand side of the road.

STOP 4 - Underwater volcanoes

On the northern side of the road here are exposures of volcanic rocks in the Lough Nafooey Group. These are green-grey in colour. It is hard to see individual crystals in these rocks since they cooled so quickly after extrusion. If you stand a little back from the exposures you should be able to see the pillow shapes of these lava flows, formed as they were forced up onto an old sea floor. In fact you are now standing on the fossilized remains of the actual floor of the Iapetus Ocean generated some 500 million years ago. The formation of pillow lavas is described in Chapter 3.

Walk up the bridle path for about 150 metres. On the right is a small disused quarry. In the southern corner, at

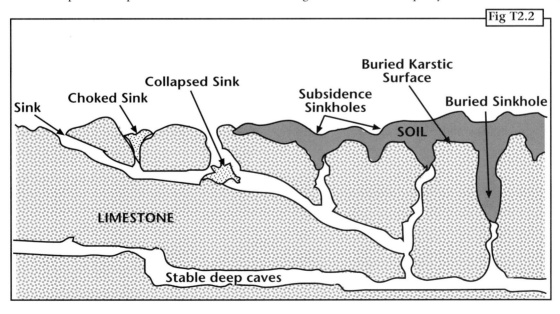

Figure T2-2. The typical features associated with karstic topography developed in the Carboniferous limestone.

floor level, is a poor exposure of black shale. This represents the sediment which settled slowly on this ancient sea floor between volcanic eruptions. This shale contains fossil graptolites which have been used to date the formation of the submarine volcanoes as earliest Ordovician, and they belong to a faunal province which developed around North America at that time. Complete specimens are rare but with the aid of a hand lens you may be able to see small shiny fragments of the skeletons of these animals preserved in the rock.

Route: Carry on westwards along the northern shore of Lough Nafooey. After about 5 kilometres there is a steep ascent with hairpin bends. Just before the top of the hill pull off to the left (there are a couple of small spaces at the side of the road).

STOP 5 - The great alluvial fans of the Ordovician

Exposed on the left-hand side of the road are Ordovician conglomerates containing some huge boulders. These sediments were deposited by massive surging floods coming from the south on the surfaces of great fan-shaped wedges of sand and gravel, alluvial fans, formed above sea level. The basin that was southern Mayo, some 460 million years ago was fringed on its southern margin by a mountainous area, probably consisting of the Dalradian of Connemara. These mountains shed their detritus towards the basin whose floor consisted of the hardened lavas of the previous stop. This erosion was accompanied by violent explosive volcanism which blew layers of ash across the surface of the sediment. When converted to rock these layers are known as tuffs. One such tuff band forms the dark ridge towering over the southern side of the road. On the northern side of the road the deep gully is the site of a fault plane running approximately east-west. There is a fine view from here across Lough Nafooey and on to the Partry Mountains.

Route: Continue westwards on this road and descend to the Maum Valley. Turn right on meeting the L100 for Leenaun. At the village fork left on the N59 for Kylemore. Drive westwards along the southern shore of Killary Harbour. This is reputed to be a fjord, a valley so overdeepened by the erosive power of the Pleistocene ice that it was drowned by the sea when the ice finally melted and sea level rose. To the north-west, across the water, can be seen the corries formed by the ice in the Mweelrea Mountains. Cross Tullyconnor Bridge and stop at Owenduff Bridge.

STOP 6 - A mix of ancient environments

Under, and around, the bridge are exposed the oldest rocks of the Silurian in this area. This rock is a breccia and is part of the Lough Mask Formation. Such a rock is very like a conglomerate but the fragments in it are very angular instead of rounded. This means that the sediment has formed virtually in situ without significant transport. The clasts in the rock are of white vein quartz and of Dalradian schist, making it an attractive rock when cut and polished. It rests unconformably on the Dalradian beneath and is thought to represent a fossilized scree deposit formed on the flanks of ancient Dalradian hills.

Walk back northwards on the road to the next small rock exposures on your left after about 50 metres. These sandstones are part of the Kilbride Formation in the Silurian sequence and overlie the breccia at the bridge. The sandstones are full of shells of the coarse-ribbed brachiopod *Eocoelia* (see Chapter 3). These brachiopods are typical of shallow-water marine environments where they were attached to rocks and other shells. On death the valves opened and separated and were concentrated by currents and storms on the old Silurian shelf. This brachiopod only existed for some 3 million years during this part of the Silurian. Other shallow water animals such as gastropods and bivalves lived in the *Eocoelia* community.

Continue walking northwards to a small bluff on the left-hand side of the road exposing conglomerates. Notice the smoothed rounded form of the boulders, showing that they had been transported some distance before being deposited. The conglomerates form part of the Lettergesh Formation of mid-Silurian age and are younger than the Eocoelia sandstones. They are thought to have been deposited in sea water on great submarine fans or perhaps at the front of a giant delta. The currents that deposited these flowed from the north.

This whole section, then, shows the Silurian marine transgression. The ancient Dalradian land surface with its river and scree deposits was gradually inundated by the sea. As the sea deepened the *Eocoelia* silts were then deposited, followed by the deeper marine gravels of the Lettergesh Formation.

Route: Continue west on the T71 past Kylemore Lough on your left and Kylemore Abbey on your right to the village of Letterfrack. Just beyond Letterfrack is the entrance to the Connemara National Park. This is worth a visit (there is a charge). Many of the aspects of the natural history of Connemara are exhibited here, including something on the geology of the area.

Continue on the road south to Clifden. Leave Clifden on the N59 eastwards to Galway. After about 6 kilometers the Twelve Pins of Connemara are visible to your left. This is Dalradian country. Many of the rocks were laid down on an ancient continental shelf some 700 million years ago. These mountains have resisted erosion because they consist largely of quartzite, a metamorphosed sandstone containing large amounts of the hard resistant mineral quartz.

Continue westwards for about 15 kilometres and turn left at the sign for Letterfrack/Lough Inagh. After 1.8 kilometres, at the top of a low rise, is a gate and rough track leading down to Derryclare Lough. Walk down the track to the shore of the lake.

STOP 7 - The Connemara Marble

Along the shore of the lake here are several old workings where the marble was extracted. In one such working, the deepest incision about 50 metres south-west of the ruined crane, can be seen a rock with purple/green/buff banding in the middle of the cut. The marbles of Connemara can take a variety of colours depending on their mineral constitution. Serpentine, for example, will impart an attractive deep green colour to the rock. The marbles here are interbanded with schists, and with rocks known as amphibolites, which weather a rusty brown in this locality. These were probably originally lava flows or small igneous intrusions into the Dalradian sediments which were subsequently metamorphosed with the growth of the minerals known as amphiboles. The marbles themselves began as limestone or dolomite deposits formed in very shallow seawater on the ancient Dalradian shelf.

The complex folding experienced by the Connemara Dalradian can be seen in several of these old workings. Unfortunately it is partly this very folding which makes the marbles difficult to quarry commercially. They occur in lenses of rock rather than a continuous outcrop and the folding makes it difficult to predict their pattern of distribution for commercial exploitation. The hills across the lake to the north-west, with their gently dipping stratification, consist of the Bennabeola Quartzite Formation, a unit which forms the Twelve Pins of Connemara.

Route: Return to the main T71 junction and turn left for Galway. Continue to Maam Cross and turn left at the crossroads here for Maum. Drive up through the pass and about 4 kilometers from Maam Cross, pull in to the left when over the summit of the pass at a small lay-by. About 300 metres to the north and north-east across the boggy ground are smoothed hummocks of rock near the foot of Leckavrea Mountain. Cross the bog with care, especially in wet weather.

STOP 8 - Free garnets

These are Dalradian rocks, here of Precambrian age. If you look carefully at some of the smoothed rock surfaces you will see large pebbles and boulders within the rock. This is the Cleggan Boulder Bed. It is the result of deposition from floating ice some 600 to 700 million years ago. This glaciation affected a large part of the earth at the time and similar glacial deposits of this age are found in Donegal (see Trail 1), Scotland and Scandinavia. Cutting across the rocks are small intrusions of granite made up of crystals of the minerals quartz (clear), feldspar (pink or cloudy grey) and mica (shiny black). These are tongues of the large Oughterard Granite to the east which have invaded these Dalradian rocks some time after their deposition. The fluids and gases associated with granite intrusion often form new minerals within the rocks which are being invaded.

Just to the south-west of these exposures is a small pit in the ground, usually partly filled with water. This is a trial cutting dug to ascertain the presence of any copper mineralization. In small tips of loose rock here are the minerals epidote (green), calcite (pink in this case), chalcopyrite (a copper sulphide with a golden metallic sheen) and garnets. The garnets are quite large but are dark brown and cloudy, so unfortunately not of gem quality.

Route: Return to Cong if desired by continuing to Maum on the L100. Turn right at the bridge and through Cornamona, follow signs to Cong.

Trail Guide Number 3 - The Clare Trail

North-western Clare offers a unique combination of spectacular mountain and coastal scenery with contrasting geology, a rich and varied archaeological heritage and, particularly during the Spring, a remarkable flora. Some 350 million years ago tropical seas with abundant animal life covered the roots of the eroded mountain landscape remaining after the Caledonian orogeny. The deposition of these limestones, which now provide the basis for the striking moonscapes of the Burren, was suddenly interrupted as the crust subsided during the early Namurian to form a basin with a centre near the present Shannon estuary. The fill of the basin was variable, ranging from, initially, black stinking muds in the Clare Shales to, later, the deltaic cycles manifest in the magnificent Cliffs of Moher.

On this trail you can appreciate the geological history of north-western Clare during the Carboniferous. You can examine the types of sediments deposited in a variety of environments during the collapse of this part of Atlantic Ireland, collect and study the fossil animal and plants inhabiting Carboniferous Clare and view the landforms developed more recently by the passage of ice and rainfall.

Ordnance Survey 1:50,000 Map Nos. 51, 57 or suitable alternative.

Route: The trail begins in Ballyvaghan. Follow the signs to Aillwee Cave.

STOP 1 - Aillwee Cave (entrance charge)
Aillwee Cave, formerly Mc Gann's Cave, was discovered in 1940. The cave penetrates the lowest part of the so called

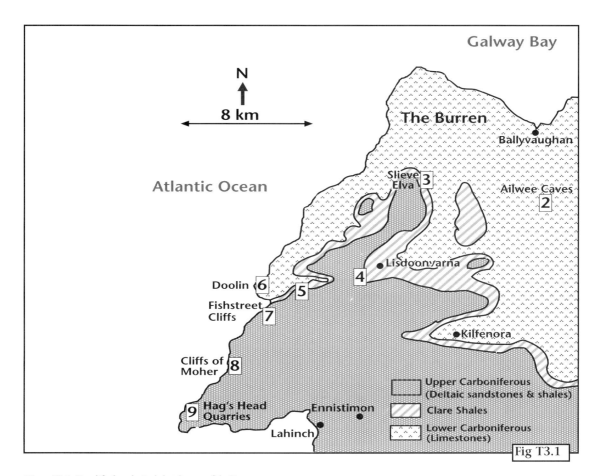

Figure T3-1. Simplified geological sketch map of the Burren area.

terraced limestone; the distinctive stepped topography of this unit contrasts with the underlying, more massive, limestone member and is exposed at and above the level of the cave. The track of the main cave follows a well developed phreatic tube and a wide range of other solution features are evident in the caverns including various stalagmites growing from the cave floor and stalactites growing downwards from the ceilings. Inside, the bones of brown bears 1500 years old have been found. Perhaps they used the cave as a hibernation den. The cave follows thick beds of limestone, some with giant brachiopods, at the base of the terraced limestones. The downward percolation of rain water was blocked by an horizon of volcanic clay, a clay wayboard, which is exposed behind the visitors' centre. The clay lies on an undulating, mamillated, surface of limestone. Similar clays in England have a fossil soil developed in the higher parts of the profile, punctuated by plant roots, showing that occasionally the limestones were exposed above sea level.

Route: On leaving the cave complex join the L51 and drive south. The road ascends part of the limestone plateau of the eastern Burren. The distinctive topography of the terraced limestones together with their karstification are obvious features of this barren landscape.

STOP 2 - The ancient people

Glenisheen and Poulnabrone, two adjacent stops, on top of the limestone pavement, offer an opportunity to examine some of the rich archaeological heritage of the Burren. At Glenisheen a wedge tomb, constructed about 4500 years ago, sits on a pavement of Carboniferous limestone. Here the terraced limestone is cut by two sets of joints. The major set is orientated approximately north and south while a second set is approximately at right angles. These joints are progressively enlarged by the dissolving effects of rain water and form sheltered habitats for the flora of the Burren. At Poulnabrone there is a classic example, much photographed, of a dolmen, constructed about 4000 years ago. These structures may have been originally covered by large mounds of stones and sods of turf. The mounds have been removed, leaving the central stone chambers of these structures. Evidence of a number of burials have been discovered here. Also soil layers were found under the supporting stones suggesting the presence of a soil cover on the Burren at the time of construction.

Route: Drive west across country to meet the main road to Lisdoonvarna, the N67. Drive north west to the lower slopes of Slieve Elva and the cave system at Poulnagollum.

STOP 3 - Giant brachiopods

Poulnagollum is one of a series of caves developed beneath the line of contact between the highest Carboniferous limestones (the Slievenaglasha Formation) and the overlying Clare Shales. In the fields adjacent to the cave system, limestones packed with crinoid ossicles crop out and a bed with large gigantoproductid brachiopods is exposed; this bed also occurs within the cave. Locally the fossils and often the matrix are silicified. Silica replacement has roughened and even sharpened the surface of the limestone. Well-preserved, silicified fossils can be extracted from these limestones by immersion for a few days in a dilute organic acid such as vinegar.

Nearby the basal part of the Clare Shales is exposed in a stream section on the lower slopes of Slieve Elva. Here and elsewhere in north-western Clare, the Clare Shales rest on the uppermost Carboniferous limestones with a slight disconformity. Deposition stopped for possibly a few million years, so there is an interval of geological time missing from the rock record. Perhaps at the edge of this developing basin the margins were initially emergent above sea level.

Route: Rejoin the N67 and drive south-west to the spa town of Lisdoonvarna, famous for its mineral springs and match-making festival. In Lisdoonvarna follow the road for Spectacle Bridge and park at the gates of the sewage works before the bridge.

STOP 4 - A fossil graveyard

A deep gorge, adjacent to the sewage works on a tributary leading into the Aille River at Spectacle Bridge, exposes an excellent section through the Clare Shales. Large calcareous bullions are sometimes crowded with the complete shells of goniatites. Following the death of these swimming molluscs their shells accumulated as fossil graveyards probably in depressions on the sea floor. Three main types of goniatite are present, *Homoceras*, *Hudsonoceras* and *Reticuloceras*. Near the top of the gorge horizons of more silty, less sulphurous sediments are packed with the shells of the scallop *Dunbarella*. These large, flat shellfish probably rested on the less toxic seabed between swims. The environments of the upper parts of the Clare Shales were much less restricted and the water was fresher and better oxygenated.

Route: From the sewage farm return a short way towards Lisdoonvarna before turning west and driving to Roadford.

STOP 5 - Fool's Gold

Here the junction between the Carboniferous limestone and the phosphatic basal beds of the Clare Shales is exposed on the west bank of the stream joining the Aille River, but only at low water. Again the lowest part of the Namurian is missing owing to a pause in deposition. Moreover the basal beds, rich in phosphate, are condensed, that is they were deposited very slowly over a relatively long period of time. Until the 1940s, phosphate was mined locally mainly for fertilizer; some of the extraction plant, now disused, can be seen opposite on the eastern side of the stream.

The section continues south in a series of low cliffs on the west bank of the stream where the black shales with sporadic calcareous bullions often packed with goniatites in full relief are exposed. Some of the fossils have been replaced by the iron sulphide, iron pyrites or fool's gold. A goniatite assemblage similar to that at the Spectacle Bridge locality occurs.

Route: From Roadford drive west through the village of Fisherstreet to Doolin Pier.

STOP 6 - Polo mints

Around Doolin pier, and to the north, pavements of the youngest Carboniferous limestone in the area, marking essentially the southern margin of the Burren, are cut by two dominant sets of joints exploited and enhanced by solution weathering. The limestone locally is packed with shoals of crinoid ossicles, resembling clusters of tiny Polo mints. Occasional brachiopods, bryozoans and corals occur. Large blocks adjacent to the slipway and around the car park display sections through shell beds crowded with large, gigantoproductid brachiopods.

The view south, however, illustrates the entire history of crustal subsidence in north-western Clare during the later Carboniferous. Standing on the limestone platform seaward of the pier and looking south, the dark cliffs in the foreground encircling the beach at Fisherstreet consist of the Clare Shales. In the middle distance the Fisherstreet Cliffs rise from the sea.

In the background the siltstones and sandstones of the overlying deltaic sequences soar abruptly out of the Atlantic to reach heights of around 200 metres.

Route: From Fisherstreet follow the road to Doonnagore Castle initially but where the road bends sharply south take the cliff track.

STOP 7 - Fisherstreet Cliffs

Great care must be taken when examining these exposures as the cliff platform is usually wet and slippery. The sedimentary rocks on and just above the platform are folded and faulted. This package of contorted strata, the Fisherstreet Slide, is about 20 metres thick; the rocks above and below the slide are undeformed. The deformation of this package occurred while the sediments were, in some cases, partly lithified. The rocks in the slide probably detached themselves from the delatic margins of the basin during periods of instability possibly encouraged by earthquakes. On their journey downslope, southwards, the strata were folded and possibly broke along faults; further disruption occurred when the jumbled package reached the base of the slope and abruptly stopped.

When a pile of poorly consolidated sediment is dumped suddenly on the sea floor the sediments settle and fluids migrate upwards through the heap. There are many dewatering structures in this section; the most spectacular are the sand volcanoes which are found on a single horizon capping the Fisherstreet Slide.

Route: Return to the road to Doonagore Castle and ascend the steep hill to the main road to the Lahinch, the L54. Drive south and turn west at the signpost for the Cliffs of Moher.

STOP 8 - Cliffs of Moher

The Cliffs of Moher offer some of the most spectacular coastal scenery in western Europe. Rising from the sea to heights of over 200 metres, they graphically display the deposits of a large delta system which periodically built out, from the west, over a basin now choked with sediment. Alternations of lighter and darker rocks stand out in the cliff faces. The darker strata are usually shales and siltstones which developed in front of the delta; they often contain a marine fauna of goniatites. The lighter rocks are sandstones associated with the development of channels on the delta, which can contain terrestrial plant material and occasional large fossil tree trunks have been reported.

A thick unit of black shale is developed above the main platform extending seaward from the edge of the car park. This unit represents a major marine transgression over the first cyclothem of deltaic deposition. The shales contain a distinctive species of goniatite, *R.* aff. *stubblefieldi*, which can be traced throughout Europe and its presence allows the very precise correlation of this marine event.

At the foot of the cliffs large blocks of rock have accumulated from a series of cliff falls. The softer shales and siltstones are continually worn away by erosion undercutting the more resistant sandstone horizons. The harder blocks are then detached along major joints and topple into the sea. Puffin Island is one of the larger fallen blocks.

Looking north-east from O'Brien's Tower about five incomplete minor cyclothems are displayed in the cliff. In the upper part, a particularly obvious light sandstone bed cuts the darker siltstones with a prominent erosional base. This shows clearly that the sandstone was originally deposited in a channel.

Route: Rejoin the L54 and continue for a short distance south before taking the first turn west after the Cliffs of Moher and continue to Johnstone's Quarry.

STOP 9 - Hag's Head Quarries

Cyclothems dominate the upper parts of the Namurian succession in western Clare. Within the Central Clare Group, above the *Reticuloceras* aff. *stubblefieldi* marine band, a variety of sediments deposited on the delta flood plain are exposed in a number of small quarries adjacent to Hag's Head. The quarries were developed to exploit the attractive Liscannor Flags. Johnstone's Quarry exposes a section through wave-rippled siltstones covered by the meandering feeding trails of an unknown mollusc or worm. These animals must have ingested tons of sediment as they systematically mined the expanses of the flood plain for nutrients. The flood plain was also incised by river channels carrying coarser material seaward. At the rear of the quarry a sandstone unit with large scale cross lamination cuts the siltstones with an erosional base. The sandstones often contain plant material transported out onto the delta top from land with a lush tropical vegetation. Fossilized leaves and stems are common in sandstone blocks scattered in the western part of the quarry.

Route: Return via the L54 to Lisdoonvarna. Continue on the L54, the coast road, through Fanore, around Black Head to Ballyvaghan. The road passes some fine scenery where coastal erosion has modified the karst developed on jointed lower Carboniferous limestones.

Trail Guide Number 4 - The Dingle Trail

The Dingle peninsula juts bravely out into the Atlantic Ocean and includes arguably the westernmost point on the Irish mainland. Its topography is dominated by Brandon Mountain, which comprises a thick sequence of Silurian to Devonian rocks reaching nearly 850 metres above sea level. Its geology includes a detailed record of sea level fluctuations in the Silurian, indications of past climate and landscape in the Devonian, and evidence of the effects of the most recent glaciers to affect the country.

Ordnance Survey 1:50,000 Map Nos. 70, 71 or suitable alternative.

Route: Start at Dingle town. Leave on the road west to Ventry and round the coast road to Slea Head. Stop at Slea Head; the nearby car park is safest.

STOP 1 - An ancient river deposit
The rocks exposed locally here are part of the Slea Head Formation, and are thought to have been deposited by ancient river systems flowing to the north-east. The pebbly sandstones of this formation show cross-stratification (see Chapter 2). These sandstones would have formed as dunes or bars within a large river. The age of these rocks is Late Silurian to Early Devonian, that is about 400 million years old. Slea Head also offers dramatic seascapes and views of the Blasket Islands to the west, which consist of Silurian and Devonian rocks generally continuous with those of the mainland.
Route: Continue north to Dunquin Harbour and park at the cliff tops above the harbour.

STOP 2 - A fall in sea level
This is a small picturesque harbour, with access down a steep concrete path. There are boats to the Blaskets in the summer. Here, at beach level, can be seen a transition in environment from the yellow-brown Silurian rocks in the

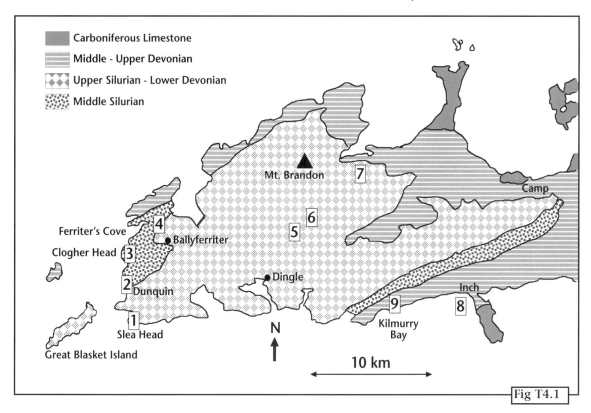

Figure T4-1. Simplified geological sketch map of the Dingle peninsula.

northern part of the section to the more reddish sandstones and mudstones to the south. The Silurian rocks are part of the Dunquin Group and were deposited in a marine environment. Similar marine rocks can be examined later. These form a transition with the red rocks of the Dingle Group to the south deposited above sea-level in rivers. Within these rocks fossilized mud cracks can be seen under the western end of the pier which demonstrate that the muds were dried out by being exposed to the air. These rocks then represent a regression, a period in the late Silurian when the sea level fell and earlier marine environments were replaced by the incursions of rivers.

Route: Drive north to Clogher Head and park at the car park.

STOP 3 - The Dingle volcanoes

The rocks exposed on the headland here are all that remains in Dingle of large Silurian volcanoes that formed on the southern margin of the Iapetus Ocean some 420 million years ago (see Chapter 3). They are mainly tuffs, the result of the explosion of volcanoes and the settlement of the clouds of ash, dust and rock fragments generated by the eruptions. Some of the tuffs contain fragments of dark, dense lava erupted at the same time. Amongst these rocks are welded tuffs, formed by the deposition of hot ash and pieces of lava which remain hot after settlement to form flattened and stretched fragments welded within the tuff. Such tuffs may be seen at Minnaumore Rock, an inland bluff some 500 metres to the south-east of the car park.

Route: Proceed north, descending on the road towards Ballyferriter. After about 2 kilometres, branch left on a minor road signposted to Dun an Oir Hotel and drive to the beach in front of the hotel at Ferriter's Cove.

STOP 4 - Silurian life

The rocks along the southern margin of the beach are part of the Ferriter's Cove Formation, almost the oldest part of the Silurian succession on Dingle. Here they consist of red and brown siltstones, thin sandstones and some tuffs. Tuffs can sometimes be distinguished by their irregular, pock-marked weathering. These sediments were laid down in a shallow marine environment near a volcanic centre; their marine origin is proved by the presence of fossils. These are quite small so it is best to look for them in any soft-weathered pockets in the rocks and use a hand lens. The fauna is diverse and includes the corals *Favosites* and *Heliolites*, the large coarse-ribbed brachiopod *Holcospirifer* (known only from Dingle) and bryozoans, bivalves, gastropods and occasional trilobites. The lenses of shells may have been concentrated in this form by Silurian storms acting on shallow-water sediments.

Some of the younger faunas in the overlying Clogher Head Formation are not so diverse. These Silurian communities may have had little time do develop on the sea floor between volcanic eruptions. Farther south-west along this section are small slump folds in the sedimentary rocks which may have been caused by the sudden arrival of a volcanic ash cloud on top of the wet sediment in the Silurian. On the north side of the beach are some quite attractive conglomerates showing the past activity of strong currents in this marine environment. In the Silurian, Dingle is thought to have been the site of volcanic islands, some distance away from the southern continent of Avalonia. It has been suggested that the main local volcanic centre was situated some 15 kilometres away from here, off what is now the west coast of the peninsula.

Route: Return to Dingle town via Ballyferriter and Ventry. Alternatively you could spend the rest of the day in this intriguing area visiting the various archaeological sites such as the Gallarus oratory near Emlagh, or the old fort at Dun an Oir in Smerwick harbour. Leave Dingle on the road north to the Connor Pass. At the top of the pass stop at the large car park.

STOP 5 - The Ice Age in Dingle

Looking north from the car park is a fine view of the Owenmore Valley. Many of the small lakes visible were dammed in their positions by dumps of boulders and sand (moraines) left as the Pleistocene ice retreated. Corries are visible in the sides of the surrounding mountains. In these the ice was generated to feed the composite glacier which scoured out the Owenmore valley. The linear ridges running along the sides of the valley are lateral moraines, sediment dumps along the side margins of the glacier.

Route: Drive north descending for about 1 kilometre where there is a small car park on the right-hand side of the road.

STOP 6 - A classic corrie
The lake just above the road here is Lough Doon (or the Pedlar's Lake). It is situated in an excellent example of a corrie. The mountain was sculpted into a hollow bowl shape by the freeze-thaw action of the Pleistocene ice. This overdeepened the corrie so that it now has a rock lip on its downstream margin. The rocks around this lip exhibit glacial striations, scratch marks on the smoothed surface of the rock formed during abrasion by boulders within the grip of the ice being scraped along the bedrock. By measuring the direction of such marks, and other features, the movement of long-vanished glaciers can be ascertained.

Route: Descend northward and stop at the bridge of Ballyduff at the bottom of the pass.

STOP7 - The deposits of the glacier
While the Pleistocene glaciers were powerful agents of erosion and sculpted much of the Irish landscape, they also transported and deposited large amounts of sediment. At the bridge here can be seen two types of deposit. At the base and top are accumulations of boulder clay, a sediment which is usually unsorted and contains a wide variety of particle sizes from boulder down to clay which was deposited directly from the glacial ice. In between the two layers are some better sorted gravels. These were probably deposited during a temporary melting phase of the glacier, when water was the transporting agent. The viscous nature of ice means it carries all material, irrespective of size, whereas water leaves the coarser material upstream and winnows out the finer material downstream. The sediment thus becomes sorted.

Route: Continue east to join the main T68 road to Tralee. At the intersection turn sharp right and proceed to Camp village. Turn left at the village up into the Slieve Mish mountains to join the L103 road on the south coast of the peninsula. At the junction with the main road turn right for Inch. Stop near the Inch Hotel at Inch shore.

STOP 8 - The vanished basement rocks
This stop is best at low water when one can walk westwards along the beach in the direction of Dingle. Exposed along the strand here are the rocks of the Inch Conglomerate Formation of Devonian age. As the name suggests these rocks are mainly conglomerates with subordinate red sandstones. The size of the boulders shows that they were deposited by strong currents. It is thought that they were deposited partly in channels on alluvial fans, where periodic flash flooding created water surges strong enough to transport the boulders. Palaeocurrent analysis has shown that these great fans probably prograded northwards at this time. The size of the boulders also indicates that they were derived from pre-existing rocks not far away in the Devonian.

If you examine the nature of the boulders you will see that they consist of a wide variety of different rock types including metamorphic rocks such as gneiss, schist and metaquartzites (see Chapter 2), as well as deformed granites, white vein quartz and some green serpentinite. This must mean that a metamorphic source was exposed nearby to the south at the time. Today there are no such rocks exposed in the vicinity, in fact the nearest such rocks are seen in Connemara far to the north and at Rosslare in the south-east of the country. These conglomerates therefore represent the composition of the old basement rocks which underlie this part of Ireland and may be fragments of the ancient sub-continent of Avalonia. This basement has since been eroded and/or covered by younger rocks.

Route: Carry on westwards towards Dingle. Just before Annascaul turn left on a minor road towards Kilmurry and Minard Castle. Park at Kilmurry Bay near the castle. Again this stop is best at low water.

STOP 9 - Desert sand dunes

The rocks exposed here are largely of the Kilmurry Formation which is Devonian in age. They consist of grey-red sandstones with occasional thin mudstones, and exhibit cross-stratification on a large scale. Whilst the beds dip at about 30 degrees to the south here, the planes of cross-stratification dip at a steeper angle. This type of large-scale structure is typical of that formed by modern desert sand dunes. The individual planes of the cross-stratification represent the advancing fronts of the giant dunes, some of these Devonian examples were probably over 30 metres high originally. They migrated over the surfaces of the alluvial fans of which the Inch conglomerates were a part.

In these sections you may be able to find sandstones with thin layers of mudstone on their tops showing fossilized mudcracks. These rocks represent the deposits of temporary rivers flowing through this desert landscape probably as flash-flood deposits related to sudden rainfall in the high source region to the south. So we can see that during the Devonian, this part of Ireland was a dry sandy terrain with little or no vegetation to stabilize the sand dunes and so prevent their continuous migration in reponse to the winds 400 million years ago.

Route: Drive on north-west to join the T68 road. At the junction turn left towards Dingle town. After about 4 kilometres take the turn to the left for Trabeg. Proceed down this minor road about 3 kilometres to Trabeg. There is a car park at the shore here.

STOP 10 - An unconformity

It is relatively rare to see an unconformity well exposed on the earth's surface (see Box 3.2). Here, south of the beach, is a classic unconformity between the Glengarrif Harbour Group and the underlying Dingle Group. Walking south to the viewing point (Figure T4.2) various formations within the Dingle Group are exposed. The Trabeg Conglomerate Formation, exposed at the southern parts of the beach, is thought to have been deposited by rivers. It contains mudcracks and cross-stratification.

Ascend from here onto the cliff top and walk to the viewing point. Looking south across the inlet from here the unconformity can be seen in the facing cliffs. At the base of the cliffs are the steeply-dipping beds of the Bull's Head Formation, originally deposited in an Early Devonian lake. These are overlain by the shallow-dipping beds of the Glengarrif Harbour Group which are Upper Devonian. The plane separating these two rock masses is the unconformity which signifies that the lower beds were first buried, deformed in the earth's crust, uplifted back to the surface and eroded. The subsequent sediments were then laid on top and finally the whole was reburied and tilted by earth movements before being uplifted again to its present position. The plane of unconformity then, represents a gap in the geological record, a time interval of perhaps 20 to 30 million years.

Route: Return to the main T68 road and turn left for Dingle town.

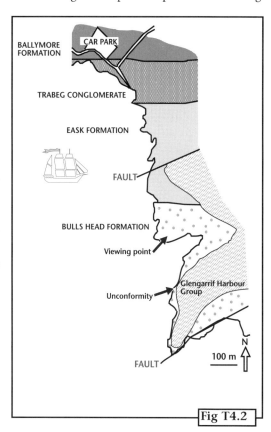

Figure T4-2. The geology of Trabeg Strand.

Trail Guide Number 5 - The Wexford Trail

The Leinster Massif occupies much of the area from just south of Dublin to the Wexford coast. The Lower Palaeozoic sequences originated in basins around the microcontinent of Avalonia. After the closure of the Iapetus Ocean these rocks were intruded by the Leinster Granite, a late Caledonian pluton trending roughly north-east to south-west and forming the core of the massif. The Lower Palaeozoic rocks are unconformably overlain by Upper Devonian breccias and conglomerates which themselves pass upwards into fossiliferous Lower Carboniferous strata.

Ordnance Survey 1:50,000 Map No. 77 or suitable alternative.

Route: The first stop can be approached from Wexford on the L128A, turning south at the junction just before Wellington Bridge. Head for Blackhall on the Cullenstown Strand.

STOP 1 - The old foundations of Ireland

Cullenstown Strand may be reached from Blackhall. At the eastern end of the strand, east of the boat harbour, metasediments of low metamorphic grade, the Cullenstown Formation, are exposed. This is an Upper Precambrian (or slightly younger) unit resting unconformably on the basement metamorphic rocks of the Rosslare Complex (see Chapter 2). In broad terms the Cullenstown Formation is the same age as the Dalradian of Connemara and Donegal. However, together with its high grade basement, it probably developed on the margins of Gondwana rather than Laurentia.

In a traverse westwards along the strand, all the main Lower Palaeozoic players crop out. Although the contacts between the main units are faults, representatives of the Cahore, Ribband and Duncannon groups are well exposed. The Ribband and Duncannon groups will be treated in more detail later. At the boat harbour the Cullenstown

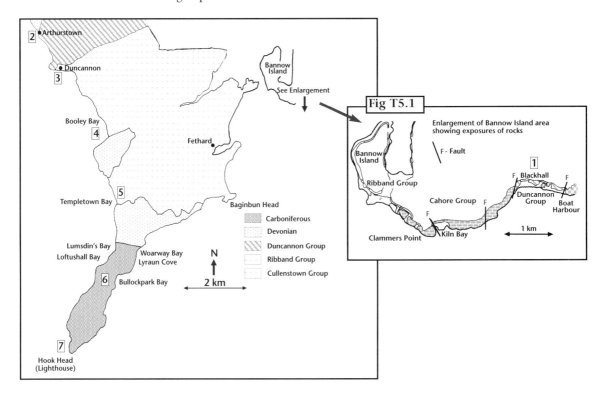

Figure T5-1. Simplified geological sketch map of a part of south-eastern Ireland.

Formation is faulted against the mainly volcanic and volcaniclastic upper Ordovician Duncannon Group. West of Blackhall, between Bridges Chamber and Kiln Bay, the lower part of the Cambrian Cahore Group is exposed. The Cahore Group is equivalent to the Bray Group which crops out mainly in the north-eastern part of the Leinster Massif. Here a series of debris-flow deposits or melanges (see Chapter 3) are interbedded with turbidites and slates. Some geologists suspect that the basement of the Rosslare Complex and Cullenstown Formation was racked by active faults shedding blocks of rock into the marginal basin where the Cahore sediments were accumulating.

In Kiln Bay and on Clammers Point the older units of the Cambrian Cahore Group are exposed. Here they consist of slates and sandstones probably deposited in a deep-water basin. If time permits, you can see where the Ribband Group is exposed around Bannow Island, where it comprises sandstones and slates, sometimes multicoloured.

Route: Return to Wellington Bridge via the L128A. Follow the L159 west.

STOP 2 - Arthurstown Bay

Park at the pier on the southern side of the bay. Here, thinly-bedded mudstones and siltstones are colour banded, and a number of folds with a well-developed cleavage are exposed. This unit, the Arthurstown Formation, is part of the Duncannon Group. This locality clearly demonstrates the relationship between the Lower and the Upper Palaeozoic sequences in the region.

The marked unconformity between the contorted Upper Ordovician Duncannon Group and the overlying upper Devonian basal breccia is exposed on the foreshore at the northern side of bay.

Route: From Arthurstown rejoin the L159 and drive east before turning south on the coast road for Duncannon.

STOP 3 - Volcanic islands

This is the type development of the Duncannon Formation, mainly volcanic and volcaniclastic rocks. The rocks of the Duncannon Group probably formed a series of islands along the northern margin of Avalonia during the mid- to late Ordovician. Locally the island shelves were colonized by abundant marine faunas usually dominated by brachiopods and trilobites whilst offshore the skeletons of graptolites accumulated. The lower part of the unit is exposed along Duncannon Strand whereas the upper parts crop out around the Old Harbour.

A wide variety of volcanic and volcaniclastic rocks are exposed along a traverse of about 100 metres around Duncannon Strand. In broad terms the sequence is ascended westwards from first a section of andesites and tuffs overlain by an ignimbrite forming a marked promontory on the eastern side of the strand. Ignimbrites are rocks formed from volcanic explosions containing large amounts of gas, the rock pumice being a typical by-product of such explosions.

These ashfall deposits are succeeded by a sequence of both laminated and massive tuffs which pass upwards into vesicular andesitic lava flows, cropping out between the western end of the strand and Duncannon Point.

Between the New Pier and the Old Harbour the middle volcaniclastic horizons crop out, dominated again by volcanic tuffs. East of the Old Harbour a conglomerate overlies the dacite volcanics which mark the base of the upper part of the sequence. North of the Old Harbour fossiliferous mudstones and shales, with Upper Ordovician graptolites, are sporadically exposed in a stream section adjacent to the coast. This is one of many faunas strung out along the entire outcrop of the Duncannon Group in eastern Ireland.

Route: From Duncannon follow the coast road south to Booley Bay, signposted Booley Strand.

STOP 4 - The sediments of ancient turbidity currents

The southern part of Booley Bay is the type development of the Booley Bay Formation, part of the Ribband Group. There is little deformation of these rocks and over 200 metres of fine-grained siliceous sediments are exposed between the northern end of the strand and the faulted contact with the Old Red Sandstone on the southern side of the section. A spectacular variety of sedimentary structures have been recorded from this section. Careful examination should reveal cross- and parallel stratification (see Box 2.3), and graded bedding in which grains become progressively smaller higher up in one bed. This represents deposition from a waning current. You may also see things called flute casts and bounce marks. These are impressions formed on a mud by a turbidity current flowing over it, are often preserved as casts on the base of the turbidite. So as well as being useful in determining the direction of ancient currents, these structures can be used to tell us whether beds are upside down or not. There are also several different types of ripples and various furrows, scours and striations. Sedimentary dykes (fossilized intrusions of sand, a bit like the sand volcanoes of Clare), disturbed laminations and shrinkage cracks are exposed at the South Booley Bay Point.

Most workers agree that the Ribband Group was deposited in deep water on the outer, or distal, parts of submarine fans, probably as dilute turbidites (see Box 2.4). No recognizable macrofossils (those visible with the naked eye) have been reported from the group, but microfossils suggest that the group ranges in age from Cambrian to early Ordovician.

Route: Follow the coast road south to Templeton Bay.

STOP 5 - Templetown Bay, adjacent to the Templars Inn.

On the shore the sedimentary rocks of the Ribband Group have been thrown into an antiform (a type of anticline) with a well-developed cleavage. Abundant flute casts indicate that the sequence is inverted.

Above the exposure in the cliff there is a striking unconformity between the Ribband Group and the overlying Upper Devonian conglomerates; the plane of the unconformity dips south. Farther south the conglomerates are exposed on the shore and were probably deposited by streams and rivers tumbling off the eroding Caledonian mountains.

Route: Continue south on the coast road until you meet the junction with the main road from Fethard which bisects the Hook Head peninsula. Most of the localities may be reached on foot from the central part of the peninsula.

STOP 6 - Hook Head peninsula

The magnificent Carboniferous succession on the Hook Head peninsula, although only encompassing the Tournaisian (Lower Carboniferous), is locally very fossiliferous. It marks a major marine transgression over the mainly Devonian Old Red Sandstone facies. The succession dips roughly south-east at about 20 degrees. The youngest rocks are thus located along the south-eastern coast of the peninsula. The lower part of the sequence is called the Porter's Gate Formation and is exposed in Lyraun Cove and Woarway Bay on the eastern side of the peninsula and also in Lumsdin's and Loftushall bays on the western side. The formation records the initial stages of the early Carboniferous marine transgression from shallow-marine sands and silts to thin shelf limestones. Phosphatic nodules and fish teeth occur throughout the upper part of the unit, together with other fossils.

The upper part of the sequence, named the Hook Head Formation, occupies most of Hook Head and is exposed on both the eastern and the western sides of the peninsula. The section is accessible from Sandeel Bay to the north. A short walk south along the coast traverses thick units of brown, cross-bedded sandstone. In Lyraun Cove the base of this unit is taken at the first 20 centimetre thick limestone exposed in the north-east facing cliff to the left of the small lime kiln when viewed from the beach. This bed is packed with productide and spiriferide brachiopod shells. A red ironstone is exposed 2 metres above the limestone. Much of the lower 60 metres of the unit consists of dark limestones and shales; the shales are highly fossiliferous with abundant brachiopods, bryozoans, corals and crinoids. A thick dolomite unit, exposed in Bullockpark Bay, is developed in the middle part of the formation. Above the dolomite the thinner bedded limestones and the less abundant shales contain relatively few fossils.

These limestones and shales were probably deposited in open marine conditions. The Hook Head seabed, particularly during deposition of the lower part of the formation, was a tangle of both low- and high-level filter feeders such as brachiopods, bryozoans and crinoids as well as the microcarnivorous corals.

STOP 7 - The Hook

On the coast, west of the lighthouse, thinly bedded limestones and marls, dipping gently south-east, are packed with bryozoans and crinoids. Large mat-like bryozoans cover a number of bedding surfaces together with long crinoid stems and brachiopods. This congested Carboniferous seafloor was tiered. The low levels were dominated by the suspension-feeding bryozoans and brachiopods whereas the crinoids grew upwards, like skyscrapers, into the higher levels of the community. Periodically the community was swamped by mud but the seabed was soon recolonized when conditions stabilised.

Trail Guide Number 6 - The Antrim Coast Trail

The Antrim coast road is one of the most spectacular coastal routes in western Europe dominated by piebald cliffs of chalk capped by the black Antrim Basalts. The Glens of Antrim punctuate the middle part of the route as deep valleys trending perpendicular to the coast and cutting deep, westwards, into the Antrim Basalts. Torr Head and Fair Head, formed by Tertiary intrusions, provide scenic cliff drives with panoramic views east on a clear day. The route ends at the spectacular Giant's Causeway and the Portrush Rock, two remarkable geological phenomena that have influenced the evolution of geological thought.

This is a long itinerary which may be split into two day trips with an overnight stop in the busy market town of Ballycastle, where boats sail for the nearby island of Rathlin. Alternatively, stops may be selected to make a hectic day excursion from either Larne or Belfast.

Ordnance Survey 1:50,000 Map Nos. 4, 5, 9, 15 or suitable alternative.

Route: The trail begins just north of the port of Larne. From the town take the A2 coast road; about 1.5 kilometres outside the town there is a parking area on the left hand side of the road more or less opposite a small stone monument to the Men of the Glynnes, who were responsible for constructing the coast road between 1832 and 1842 as a famine-relief project. Descend to the shore here in the Waterloo area and walk south towards Larne along the promenade to the Chaine Memorial Park.

STOP 1 - Fish for tea

In a short traverse from south of the Chaine Park to the Waterloo Cottages we climb the stratigraphy from the hot deserts of the Triassic through a major marine transgression, evident throughout Europe, culminating in the warm marine environments of the Early Jurassic. The entire section is only about 1 kilometre long yet we cover about 30 million years of geological time.

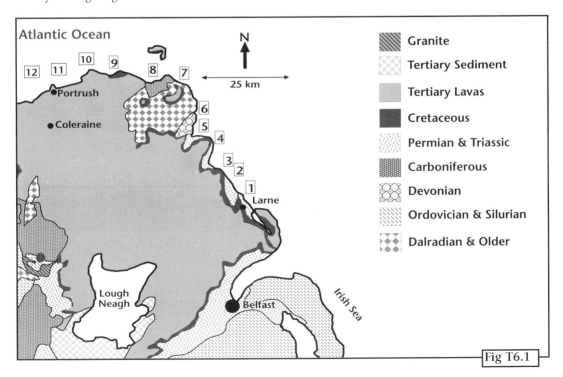

Figure T6-1. Simplified geological sketch map of north-eastern Antrim

About 500 metres south of the park and play area of the Chaine Memorial Park, red marls with green reduction blotches are exposed along the shore below the high tide mark. These red Keuper Marls were probably deposited in shallow warm playas (temporary lakes in arid climates) in a Late Triassic desert. The green blotches may be the result of local reduction of the oxidized iron minerals by fragments of organic material in the sediment. Lower in the Keuper Marls evaporite minerals such as anhydrite and gypsum occur, which is suggestive of intense evaporation in these hot mineral-saturated pools.

Walking back towards the Chaine Park you will see a series of low cliffs on the landward side of the promenade which consist of pale grey and green mudstones overlying the red marls. These sediments are included in a unit called the Tea Green Marls and contain some fish scales and the occasional remains of the small bivalved crustacean *Euestheria*, characteristic of Upper Triassic rocks elsewhere.

In a cliff cutting on the southern edge of the park area, black marls and limestones are exposed. This is a very important stratigraphical unit called the Rhaetic which includes the youngest rocks of the Triassic System and marks the start of a major marine transgression which swept across Europe. The Rhaetic contains shallow-water marine fossils such as the distinctive bivalves *Procardia* and *Rhaetavicula* as well as fish teeth, occasionally forming bone beds packed with phosphatic material.

North of the park, adjacent to the area of the Waterloo Cottages and along the shore, the basal beds of the Jurassic crop out as a sequence of dark limestones and mudstones.

Route: Follow the A2 north; initially the landward road margin is marked by walls of the Cretaceous White Limestone (chalk), but northward the overlying dark Tertiary Antrim Basalts crop out along the roadside. The basalts are penetrated by the Black Cave Tunnel on the southern side of Drains Bay. Fossiliferous blocks of Lias are often washed up along the bay. Continue around Drains Bay to Ballygalley Head.

STOP 2 - A volcanic feeder
Ballygalley Head consists of a Tertiary dolerite intrusion, nearly 100 metres high, cutting the White Limestone and characterized by columnar cooling joints. The contact of this intrusion with the chalk is exposed in the roadside quarry at the headland and locally in the contact zone a chalky agglomerate (conglomerate of volcanic material) is developed. This was probably a feeder for one of the great Tertiary volcanoes of County Antrim.

Looking northwards towards Garron Point the stepped appearance of the cliff line is particularly obvious. Large blocks of near horizontal White Limestone capped by the Antrim Basalts have detached, slipped seaward and rotated on a poorly consolidated base of Lias mudstones. The cliff topography is dominated by these landslipped blocks and is as spectacular as the better-documented structures along the Dorset coast, near Lyme Regis, in England.

Route: Continue north on the A2 towards Glenarm at the head of the first of the nine Glens of Antrim.

STOP 3 - The Irish Chalk
A number of roadside quarries, just before the village of Glenarm, some abandoned and some mining for carbonate fill, expose excellent sections through the White Limestone or the Irish development of the Cretaceous chalk, which makes up the White Cliffs of Dover in England. The White Limestone is hard and impervious; the constituent grains have been recrystallized and voids are filled with coarse crystals of calcite. Consequently the rock is much harder than the English chalk of the Downs and the Dover cliffs. There is abundant evidence of pressure solution including the stylolites that are developed throughout the sequence (zig-zag dark planes in the rock along which much limestone has been dissolved during burial). However there are nodular and tabular flint horizons and many of the same fossils found in the English chalk, although they are much harder to collect.

Route: Continue on the A2 through the beautiful arch, constructed from locally quarried blocks of the White Limestone, at the entrance to the village of Carnlough. The harbour walls have been similarly constructed with equidimensional blocks of the hard White Limestone. Drive up the coast to Garron Point and leave vehicles in the car park south of the point.

STOP 4 - Giant landslips

A number of landslipped blocks are developed around the point. Each block has an internal stratigraphy of originally near-horizontal White Limestone overlain by Antrim Basalts. The blocks have slid along a substrate of soft Lias mudstones. Some of these mudstones are very fossiliferous around the shore, but recently the Department of the Environment has covered them with concrete to prevent further sliding.

Route: Continue along the A2 coast road around Garron Point to Waterfoot at the head of Glenariff.

STOP 5 - Wadis in a Triassic desert

Thick units of Triassic conglomerates and sandstones, part of the so-called New Red Sandstone, are exposed in a series of low cliffs just north of Waterfoot and around Red Bay. The conglomerates contain pebbles of Dalradian quartzites and mica schists together with recycled clasts of quartzite from the Old Red Sandstone conglomerates and various Caledonian igneous rocks.

Equivalent rocks at Scrabo Quarry, County Down, have yielded the footprints of the reptile *Chirotherium lomasi*. Environments were arid and hot with the conglomerates deposited by sudden flash floods in an otherwise sandy desert mileau. These early reptiles were probably attracted to wadis which may have occasionally held some standing water between sudden flushing by flash floods. A raised beach approximately 8 metres above sea level is apparent here at the northern end of Red Bay.

Route: Drive through Cushendall on the A2 but deviate to Knocknacarry and then on to Cushendun. Before the bridge turn sharp right at the Cushendun Hotel and continue along the side of the small harbour to the Bay Hotel car park.

STOP 6 - Ancient river deposits

Along the coast and in sea caves south of the Bay Hotel, thick red conglomerates are exposed. These sediments are included in a rock unit called the Old Red Sandstone and probably represent the lower part of a thick, mainly river-borne, sequence of sediments which was shed from the rising Caledonian mountains in the Devonian Period. The conglomerates are packed with large pebbles of quartzite and occur with red sandstones, siltstones and mudstones with a range of sedimentary structures including cross stratification and mud cracks. During the Devonian a spine of older Dalradian rock to the north, the Highland Border Ridge, marked the northern margin of this small intermontane basin.

Route: From Cushendun take the scenic cliff drive, signposted 'Torr Head'.
From Torr Head there are views, looking north-east, of the Mull of Kintyre, the south coast of Arran and the conical intrusion of riebeckite microgranite of Ailsa Craig, famous for curling stones. To the north, Rathlin Island composed of White Limestone topped by dark Antrim Basalts, is cut by a set of north to south faults.

Drive down to the shore at Murlough Bay and follow the unmetalled track to a parking place adjacent to an old lime kiln. Continue along the track on foot, through the gate and past the small gate cottage to the white shore cottage near the slipway.

STOP 7 - Raised beach and a lot more

Here there is a very diverse menu of geology in a relatively small area; to cover the entire ground in detail many groups devote a full day to the section. On the shore Dalradian schists to the south are faulted against Carboniferous sandstones overlain by pillow lavas to the north. A raised beach is developed here, and some of the pillows were gouged out to form small sea caves.

Ascending the cliffs above the shore the Triassic is represented by over 30 metres of vivid red sedimentary rocks. The lower parts are coarse-grained conglomerates with pebbles of quartz and schist, in broad terms similar to the Triassic sediments outside Waterfoot and around Red Bay (Stop 5). Higher strata are marls like the Keuper Marls exposed near Chaine Park (Stop 1).

The basal Hibernian Greensand rests unconformtably on the Trias here. This conglomerate contains clasts of quartz pebbles and the mineral glauconite, which imparts a green colour. This mineral grows only in sediments in a normal marine environment with relatively low rates of sedimentaion. It also contains ammonites derived by the erosion of the older Lias beds, such as *Dactylioceras*. The White Limestone above contains several sponge beds with *Aphrocallistes* and *Ventriculites*.

The Fair Head Sill, a Tertiary intrusion of dolerite, dominates the northern end of the bay, rising up from the coast where it intrudes the Carboniferous rocks.

Route: Drive north and rejoin the A2 at Ballyvoy. Continue to the coast. The section extends east from Ballycastle Bay to Fair Head.

STOP 8 - The plants from the swamps

East of Ballycastle the Carboniferous sedimentary rocks, overlying the volcanics seen at Murlough Bay, are well exposed in a series of cliffs between Ballycastle Bay and Fair Head. The Middle Carboniferous was deposited as cyclothems quite like the younger Carboniferous rocks of western Clare (see Chapter 5). Marine limestones with mainly brachiopods are overlain by sandy horizons with thin coal seams at the summit of each cycle, deposited as the sea level fell. There are three major marine bands exposed along this coastal strip. The Carrickmore Marine Band, the lowest, crops out above the waterfall at Carrickmore itself. The so-called Main Limestone at North Star has a rich brachiopod fauna dominated by *Gigantoproductus* and *Spirifer*, together with *Buxtonia* and *Productus* and the overlying shales are packed with the shells of the brachiopod *Chonetes*. The McGildowney Marine Band, the highest, crops out at Pollard. A number of coal seams, all that remain of the extensive swamps of the mid Carboniferous, are exposed along the section and occasionally compressed plant fossils such as *Calamites*, *Lepidodendron* and *Stigmaria* are preserved.

Route: From Carrickmore continue into the busy market town of Ballycastle. This makes an ideal overnight stop. Alternatively you can return to Belfast via the A44 through Armoy to its junction with the A26, meeting the M2 for Belfast just north of Antrim. To continue on the trail take the B15 for Ballintoy and head for the coast at Ballintoy Harbour.

STOP 9 - White Park Bay

Access is from Ballintoy Harbour entering the bay from the east. Although the Lias crops out at the eastern entrance to the bay the exposures may be covered with sand and may require excavation. Grey mudstones of early Liassic age have a relatively diverse fauna of ammonites, gastropods and bivalves. The Lias is also exposed in the most easterly stream reaching the beach in the bay itself. The green glauconitic conglomerates and sands of the Hibernian Greensand are exposed at about the level of the beach sand and overlie the Lias with a marked unconformity. The conglomerates contain derived fossils from the Lias beneath. The overlying White Limestone forms cliffs around the bay and much of the Cretaceous sequence is within landslipped blocks again, as at Garron Point, controlled by the soft underlying Lias mudstones.

On the western side of the bay the White Limestone is faulted against the overlying Antrim Basalts. In the middle part of the beach evidence of a tumulus together with polished flints found on the beach suggest the presence of late Bronze Age and Iron Age settlement.

Route: Rejoin the B15 and continue to the junction with the A2. Branch left on the B146 for the Giant's Causeway. There is a large car park adjacent to the interpretative centre owned by the National Trust.

STOP 10 - Giant's Causeway: the work of Finn MacCool?

Legend has it that the spectacular columnar-jointed basalt platforms at the Giant's Causeway were built by the giant Finn MacCool to enable an attack to be made on a rival giant, domiciled on the Hebridean island of Staffa, where similarly jointed lava is extravagantly developed in Fingal's Cave. Several days could be spent along the Causeway and the centre has a large variety of maps and guides. It also shows films and videos. This stop highlights a few selected features of the Antrim Basalts at this historic locality.

Two main groups of flows form this part of the coast, the lower and middle basalts are separated by a striking red interbasaltic horizon. The lower basalts are exposed on the western side of the steep path that runs from the interpretative centre to the shore and the interbasaltic horizon is sporadically exposed to the east of the path. At the bend in the first downhill leg of the path recent spheroidal weathering of the lower basalts is marked by the typical onion-skin shapes.

On the Grand Causeway itself, allegedly over 35,000 polygonal basalt columns are separated from each other by a mosaic of cooling joints. The columns are inclined since the middle basalts were extruded as a magma pond. As the lava around the margins cooled, the axes of the columns developed perpendicular to the sloping lake bed. In the cliffs south-east of the causeway the Giant's Organ with its battery of organ pipes of columnar jointed middle basalt is clearly visible.

Looking west from the Grand Causeway the junction between the lower and upper Basalts is clearly marked by a vivid red horizon of lateritic clay, the interbasaltic bed. Here, between the eruptions, intense tropical weathering occurred in the hot Tertiary climate, much the same as the chemical weathering prevalent in parts of Africa today.

Route: From the Causeway rejoin the A2 for Portrush. The route passes through the town of Bushmills, where the whiskey distillery holds the world's oldest licence to distill spirits; it has an interpretative centre and free samples are part of the tour.

STOP 11 - Agglomerate stack and a lively banquet

Dunluce Castle, 3 kilometers west of Bushmills, was built in the thirteenth century on a high stack of basaltic agglomerate. Much of the castle is said to have collapsed and fallen into the sea during a presumably lively banquet in 1693. About 50 metres east of the Dunluce Burn the junction of the vent with the country rock of White Limestone is seen, capped by Antrim Basalts.

Route: Continue from Dunluce Castle along the A2 into the busy seaside resort of Portrush.
In Portrush drive to Landsdown Crescent, a row of Victorian guest houses facing the eastern foreshore on the Portrush Peninsula. Park in the large car park opposite the coast guard station.

STOP 12 - Baked fossils and controversy

The critical section here is exposed on either side of the Shelter, extending for nearly a kilometre from Reviggerly Point in the north to the Blue Pool in the south. On the foreshore an olivine dolerite sill, over 30 metres thick, has intruded the Lower Jurassic. These Liassic sediments were metamorphosed to hornfels and porcellanites, that is sediments hardened through baking by the hot intrusion. The porcellanite, known as the Portrush Rock, is fossiliferous, with relatively abundant ammonites preserved in the baked matrix. Neptunists in the eighteenth century, intepreting the porcellanites as an igneous rock, suggested that the Portrush Rock with its marine fossils provided convincing evidence of an aqueous origin for all rock types. Their adversaries, the vulcanists, believed that all rocks were crystallized from magma. This tough rock was used in the manufacture of axes by the earliest farming communities (Neolithic). This is a World Heritage Site and visitors are asked not to remove any fossils or pieces of rock.

Route: From Portrush return to Belfast is across the relatively monotonous top of the Antrim Plateau. Take the A29 for Coleraine and follow the outer ring clockwise for the A26. The A26 can be followed to just north of Antrim where it joins the M2 for Belfast. Alternatively the Donegal trail is within striking distance after an overnight stay in Portrush or in its twin harbour town to the west, Portstewart.

A glossary of some geological terms

A glossary of some geological terms

A

Accretionary prism. A thick wedge of sediments caught between two colliding plates.
Acidic. When applied to igneous rocks it means a quartz content of more than 10%.
Archosaur. A group of reptilian vertebrates containing the birds, dinosaurs and crocodiles.

B

Basalt. A fine-grained basic igneous rock usually formed from lava.
Basic. When applied to igneous rocks it signifies that no quartz is present.
Belemnite. A fossil mollusc with an internal cigar-shaped skeleton, related to living squids; nektonic.
Benthos. Life on the seabed.
Boulder clay. An unsorted mixture of sediment deposited from a glacier.
Brachiopod. A twin-shelled invertebrate with a distinctive hairy arm-like feeding organ or lophophore; benthonic.
Breccia. A rock consisting of angular fragments of older rocks.
Bryozoan. A colonial organism related to the brachiopods, minute individuals equipped with lophophores; benthonic.
Bullion. Type of calcareous nodule or concretion, commonly packed with goniatites.

C

Caliche. A deposit of calcium carbonate formed near the top of a sediment by evaporation of fluids.
Chalcopyrite. An important ore of copper in the form of copper and iron sulphide.
Chert. A hard sedimentary rock made of minutely crystalline silica.
Chondrite. A meteorite consisting of stony material with globules of minerals.
Coccolith. Calcareous hard plate of microscopic planktonic algae forming the main constituent of chalk.
Conglomerate. A sedimentary rock consisting of rounded fragments of older rocks.
Corrie. Also known as cirque (French) or cwm (Welsh), it is a hollow depression in a hillside created by the scouring action at the origin of a glacier.
Crinoid. Stalked echinoderm with arms to assist feeding, related to the sea urchins, forming the higher tiers of benthic communities.
Cross-lamination. Small laminae formed at an angle to sedimentary layering by the migration of the fronts of ripples; cross-bedding is the large-scale version.

D

Dip. The angle that a plane such as a sedimentary bed makes with the horizontal.
Dolerite. An intrusive medium-grained basic igneous rock.
Dolomite. A limestone containing more than 15% magnesium.
Drumlin. A smoothed mound of glacial deposits moulded by the passage of ice over it.
Dyke. A vertical injected sheet of igneous rock.

E

Echinoid. A mobile type of echinoderm with a globular test and moveable spines, including the sea urchins and sand dollars; benthonic.
Endemic. Restricted geographically to a specific area, region or province.
Epidote. A mineral, usually green, containing calcium, iron, aluminium and silica.
Esker. A sinuous deposit of sands and gravels deposited by rivers flowing within glaciers.
Evaporite. A mineral or rock formed by chemical precipitation from evaporating solutions.

F

Fault. A fracture in a rock mass along which there has been some movement.

Feldspar. A group of common minerals containing various combinations of silica with potassium, aluminium, sodium and calcium.

Flint. A hard sedimentary rock similar to chert but having a different type of fracture.

Flute cast. A small scour formed by the passage of a strong current over mud and filled in by later sediment.

Folds. Contortions of strata formed by crustal forces.

Fool's gold. Iron pyrites, a sulphide of iron, with a brassy yellow appearance.

Foraminifera. Microscopic, unicellular animals, usually with calcareous tests; planktonic and benthonic.

G

Gabbro. A coarse-grained basic igneous rock intruded at some depth in the crust.

Gastropod. A group of molluscs with a well-developed foot, siphons and usually a coiled shell; mainly benthonic.

Geophysics. A branch of geology dealing with the structure of the crust and underlying zones of the earth.

Glauconite. A bright green mineral containing aluminium, iron, magnesium and silica, most commonly found in marine sediments.

Gneiss. A banded metamorphic rock formed at high temperature and pressure.

Goniatite. Extinct cephalopod mollusc with coiled, chambered shell divided by simple, zig-zag partitions; nektonic.

Granite. A coarse-grained acidic igneous rock, commonly containing quartz, feldspars and micas, intruded deep in the earth's crust.

Graptolite. Small, stick-like extinct colonial animal related to the chordates; mainly nektonic and planktonic.

H

Hanging valley. A small glacial tributary valley left high above a main glacial valley because of differing rates of glacial erosion.

Hornfels. A rock which has been baked by contact with a hot igneous body.

I

Igneous. A type of rock crystallized from molten magma.

Ignimbrite. Sometimes called welded tuffs, they are formed by deposition from volcanic clouds so that the hot fragments become welded together.

Intrusion. A mass of igneous rock which has been injected into older rocks.

Iridium. An element which is common in meteorites and the core of the earth but not in the crust.

J

Joint. A crack or fracture in a rock mass along which there has been no movement.

K

Karst. A very irregular land surface caused by the dissolution of limestone.

Kerogen. A solid mass of organic material which can be a precursor to the formation of petroleum.

L

Laterite. A deposit formed by the dissolving of more soluble minerals in a tropical environment, leaving behind a material very rich in iron.

Lignite. Upon shallow burial peat converts to lignite; buried more deeply it converts to coal.

Limestone. A sedimentary rock consisting largely of calcium carbonate.

M

Magma. A molten material present within the crust, or beneath, which forms igneous rocks when it cools.
Marble. A limestone which has been metamorphosed by high temperature and pressure.
Marl. A sediment or rock made of calcium-rich clay.
Melange. A rock consisting of a chaotic mixture of large and small blocks of older rock set in a fine-grained matrix.
Mesolithic. The Middle Stone Age, in Ireland from about 9000 to 6000 years ago.
Metamorphic. A rock whose original form has been changed by subsequent heat and pressure.
Mica. A mineral found as flat semi-transparent plates containing various elements combined with silica.
Mid-ocean ridge. A linear volcanic ridge present in many oceans along which two plates are moving apart allowing the magma to erupt onto the sea floor.

N

Nautiloid. Cephalopod mollusc with a chambered shell divided by simple partitions; nektonic.
Nektonic. Active swimming aquatic animals.
Neolithic. The New Stone Age, in Ireland from about 6000 to 4500 years ago.

O

Obduction. The collision of a continental plate with that of an ocean may sometimes result in oceanic rocks being emplaced on top of the continent.
Olivine. A type of basic mineral common in basic igneous rocks.
Orogeny. An episode of earth history during which a mountain belt is formed.
Ossicles. Small discs of calcite, perforated centrally like a polo mint, comprising the stem and arms of a crinoid.

P

Palaeocurrent. A current which was responsible for transporting sediment in the past. Its direction may be measured by examining sedimentary structures in rocks.
Palaeontology. The study of fossilized organisms.
Pelite. A fine-grained sedimentary rock which has been metamorphosed.
Phreatic. A descriptive term for ground water present below the lowest level of the water table in an area.
Plate. A division of the earth's crust which may move in relation to others.
Pluton. A large mass of igneous rock.
Porcellanite. A fine-grained sedimentary rock from Antrim which was baked by an igneous intrusion.

Q

Quartz. An oxide of silica forming a hard mineral with a variety of colours but commonly white or transparent.
Quartzite. A sandstone that has been metamorphosed (also metaquartzite).

R

Regression. A drop in sea level with respect to the land.
Reservoir rock. A porous rock which contains petroleum in its pore spaces.
Rugose. Group of Palaeozoic corals including both solitary and colonial growth modes with prominent vertical radial partitions or septa.

S

Schist. A metamorphic rock with a parallell arrangement of minerals in layers.
Sedimentary rock. A rock formed from sediment.
Serpentine. A mineral containing magnesium and silica often as an alteration product of olivine.

Sill. An igneous body injected as a horizontal sheet into older rocks.
Source rock. A rock in which petroleum is created during burial.
Stromatolite. Concentric carbonate laminations constructed by algae and cyanobacteria; common in mid - late Precambrian rocks.
Stylolite. A zig-zag feature in limestones along which parts of the rock have been dissolved by water.
Subduction. The descent of an oceanic plate beneath that of a continent during collision.
Suspect terrane. A part of the earth's crust whose geological history cannot be matched with adjacent parts and is therefore suspected of having travelled some distance to its present location.
Suture. The line along which two colliding plates have finally come to rest.

T

Tabulate. A group of Palaeozoic colonial corals usually lacking septa.
Tephra. Fragments ejected from the vent of a volcano.
Tillite. A rock formed from till, the deposit of a glacier.
Transgression. A rise in sea level resulting in the flooding of a land surface.
Trap rock. A non-permeable rock which prevents the escape of petroleum from a reservoir rock.
Trilobite. An extinct arthropod with a head, thorax and abdomen, common in Palaeozoic rocks; mainly benthonic.
Tuff. A rock formed from volcanic ash.
Turbidite. A rock formed from sediment transported by a turbidity current.

U

U-shaped valley. A valley whose shape has been created by the scouring action of ice.
Ultrabasic. Igneous rocks with virtually no quartz or feldspar.
Unconformity. A plane along which older deformed rocks are in contact with younger, less deformed, ones.

V

Volcanic arc. An arc-shaped string of volcanoes formed above the leading edge of a subducting plate.
Volcaniclastic rock. A rock formed by material directly derived from a volcano.

References

We include here a short selection of publications which deal with various aspects of the geology of Ireland. We only include the more recent and general works which often contain references themselves to earlier information. Some are more readable than others and many require some background in geology. All scientific studies contain an element of subjectivity so, as they say, do not believe everything you read in the papers!

General

There are a great number of general books dealing with introductory geology. The following refer specifically to Ireland.

Holland, C. H. (editor). 1981. *A Geology of Ireland*. Scottish Academic Press, Edinburgh. A collection of chapters written by geological specialists and largely suitable for people with an educational grounding in geology.

Mitchell, G. F. 1976. *The Irish Landscape*. Collins, London. A book by a great natural historian dealing largely with the Pleistocene and later development of Ireland. A short introduction to the geology is included.

Mitchell, G. F. 1986. *Shell Guide to Reading the Irish Landscape*. Country House Publishers, Dublin. An updated version of the above with some colour plates.

Whittow, J.B. 1974. *Geology and Scenery in Ireland*. Penguin, Harmondsworth. A county by county discussion of the scenery of the country with a geological bias. Now somewhat dated.

Chapter 1

Duff, D. 1993. *Holmes' Principles of Physical Geology*. Chapman & Hall, London.

Press, F. & Siever, R. 1982. *Earth*. W.H. Freeman & Co San Francisco.

Chapter 2

Elias, E.M., MacIntyre, R.M. & Leake, B.E. 1988. *The cooling history of Connemara, western Ireland, from K-Ar and Sb-Sr studies.* Journal of the Geological Society of London, 145, 649-60.

Max, M.D. & Long, C.B. 1985. *Pre-Caledonian basement in Ireland and its cover relationships.* Geological Journal, 20, 341-66.

Murphy, F.C. 1990. *Basement-Cover relationships of a reactivated Cadomian mylonite zone: Rosslare Complex, S.E. Ireland.* Special Publication of the Geological Society of London, 51, 329-39.

Pitcher, W.S. & Berger, A.R. 1972. *The Geology of Donegal: A study of granite emplacement and unroofing.* John Wiley & Sons, Chichester.

Roddick, C. & Max, M.D. 1983. *A Laxfordian age from the Inishtrahull Platform, Co. Donegal.* Scottish Journal of Geology, 19, 97-102.

Williams, D.M. & Rice, A.H.N. 1989. *Low-angle extensional faulting and the emplacement of the Connemara Dalradian, Ireland.* Tectonics, 8, 417-28.

Winchester, J.A. & Max, M.D. 1987. *The Pre-Caledonian Inishkea Division of northwest Co. Mayo, Ireland: its geochemistry and probable stratigraphic position.* Geological Journal, 22, 309-31.

Yardley, B.W.D, Barber, J.P. & Gray, J.R. 1987. *The metamorphism of the Dalradian rocks of western Ireland and its relation to tectonic setting*. Philosophical Transactions of the Royal Society of London, Series A, 321, 243-70.

Chapter 3

Dewey, J.F. & Shackleton, R.M. 1984. *A model for the evolution of the Grampian tract in the early Caledonides and Appalachians.* Nature, 312, 115-21.

Graham, J.R., Leake, B.E. & Ryan, P.D. 1989. *The Geology of South Mayo, western Ireland (with map).* Scottish Academic Press, Edinburgh.

Harper, D.A.T & Parkes, M.A. 1989. *Palaeontological constraints on the definition and development of Irish Caledonian terranes.* Journal of the Geological Society of London, 146, 413-15.

Hutton, D.H.W. 1987. *Strike-slip terranes and a model for the evolution of the British and Irish Caledonides.* Geological Magazine, 124, 405-25.

Murphy, F.C. and others. 1991. *An appraisal of Caledonian suspect terranes in Ireland.* Irish Journal of Earth Sciences, 11, 11-41.

Williams, D.M. 1990. *Evolution of Ordovician terranes in western Ireland and their possible Scottish equivalents.* Transactions of the Royal Society of Edinburgh, Earth Sciences, 81, 23-9.

Chapter 4

Clayton,G., Graham,J.R., Higgs,K., Holland,C.H. & Naylor,D. 1980. *Devonian rocks in Ireland: a review.* Journal of Earth Sciences Royal Dublin Society, 2, 161-83.

Gardiner, P.R.R. & MacCarthy, I.A.J. 1981. *The Late Palaeozoic evolution of southern Ireland in the context of tectonic basins and their transatlantic significance.* Geology of the North Atlantic borderlands (Editors. J.W.Kerr & A.J. Fergusson). Canadian Society of Petroleum Geologists, Memoir 7, 683-725.

Stossel, I. 1995. *The discovery of a new Devonian tetrapod trackway in SW Ireland.* Journal of the Geological Society of London 152, 407-413.

Todd, S.P., Williams, B.P.J. & Hancock, P.L. 1988. *Lithostratigraphy and structure of the Old Red Sandstone of the northern Dingle Peninsula, Co. Kerry, SW Ireland.* Geological Journal, 23, 107-20.

Chapter 5

Collinson, J.D., Martinsen, O., Bakken, B. & Kloster, A. 1991. *Early fill of the Western Irish Namurian Basin: a complex relationship between turbidites and deltas.* Basin Research, 3, 223-42.

Drew, D. 1989. *New caves in the Burren?* Irish Speleology 13, 16-9.

Feehan, J. 1991. *The rocks and landforms of the Burren.* In O'Connell, J.W. & Korff (editors) The Book of the Burren, Tir Eolas, Kinvara. pp. 14-23.

Gill, W.D. 1979. *Syndepositional sliding and slumping in the West Clare Namurian Basin, Ireland.* Special paper of the Geological Survey of Ireland, 4, 31 pp.

Martinsen, O.J. & Bakken, B. 1990. *Extensional and compressional zones in slumps and slides - examples from the Namurian of County Clare, Ireland.* Journal of the Geological Society of London, 146, 153-64.

Rider, M.H. 1974. *The Namurian of County Clare, Ireland.* Proceedings of the Royal Irish Academy, 74B, 125-42.

Sleeman, A.G., Johnston, I.S., Naylor,D. & Sevastopulo,G.D. 1974. *The stratigraphy of the Carboniferous rocks of Hook Head, Co. Wexford.* Proceedings of the Royal Irish Academy, 74B, 227-43.

Chapter 6

Mitchell, G.F. 1980.   *The search for the Tertiary in Ireland.*   Journal of Earth Sciences, Royal Dublin Society, 3, 13-33.

Naylor, D. 1992.   *The post-Variscan history of Ireland*. In, Parnell,J. (editor) Basins on the Atlantic Seaboard. Geological Society Special Publication, 62, 255-75.

Parnell, J., Shukla, B. & Meighan, I.G. 1989.   *The lignite and associated sediments of the Lough Neagh Basin.* Irish Journal of Earth Sciences, 10, 67-88.

Shannon, P.M. 1991.   *The development of the Irish offshore sedimentary basins.* Journal of the Geological Society of London, 148, 181-89.

Wilson, H.E. 1972.   *Regional Geology of Northern Ireland.* Her Majesty's Stationary Office, Belfast.

Ziegler, P.A. 1990.   *Geological Atlas of Western and Central Europe.* 2nd. edition. Shell Internationale Petroleum Maatschappij B.V. Netherlands.

Chapter 7

Bellamy, D. 1986.   *The Wild Boglands - Bellamy's Ireland.* Country House Publishers, Dublin.

Carter, R.W.G., Devoy, R.J.N. & Shaw, J. 1989.   *Late Holocene sea levels in Ireland.* Journal of Quaternary Science, 4, 7-24.

Devoy, R.J. 1985. *The problem of a Late Quaternary landbridge between Britain and Ireland.* Quaternary Science Reviews, 4, 43-58.

Edwards, K.J. & Warren, W.P. (editors) 1985.   *The Quaternary History of Ireland.* Academic Press, London.

McCabe, A.M. & Dardis, G.F. 1989.   *A geological view of drumlins in Ireland.* Quaternary Science Reviews, 8, 169-77.

McCabe, A.M., Dardis, G.F. & Hanvey, P.M. 1987.   *Sedimentation at the margins of a Late-Pleistocene ice-lobe terminating in shallow marine environments, Dundalk Bay, eastern Ireland.* Sedimentology, 34, 473-93.